Textile Science and Clothing Technology

Series Editor

Subramanian Senthilkannan Muthu, SgT Group & API, Hong Kong, Kowloon, Hong Kong

This series aims to broadly cover all the aspects related to textiles science and technology and clothing science and technology. Below are the areas fall under the aims and scope of this series, but not limited to: Production and properties of various natural and synthetic fibres; Production and properties of different yarns, fabrics and apparels; Manufacturing aspects of textiles and clothing; Modelling and Simulation aspects related to textiles and clothing; Production and properties of Nonwovens; Evaluation/testing of various properties of textiles and clothing products; Supply chain management of textiles and clothing; Aspects related to Clothing Science such as comfort; Functional aspects and evaluation of textiles; Textile biomaterials and bioengineering; Nano, micro, smart, sport and intelligent textiles; Various aspects of industrial and technical applications of textiles and clothing; Apparel manufacturing and engineering; New developments and applications pertaining to textiles and clothing materials and their manufacturing methods; Textile design aspects; Sustainable fashion and textiles; Green Textiles and Eco-Fashion; Sustainability aspects of textiles and clothing; Environmental assessments of textiles and clothing supply chain; Green Composites; Sustainable Luxury and Sustainable Consumption; Waste Management in Textiles; Sustainability Standards and Green labels; Social and Economic Sustainability of Textiles and Clothing.

More information about this series at http://www.springer.com/series/13111

Subramanian Senthilkannan Muthu
Editor

Environmental Footprints of Recycled Polyester

 Springer

Editor
Subramanian Senthilkannan Muthu
SgT Group & API
Hong Kong, Kowloon, Hong Kong

ISSN 2197-9863 ISSN 2197-9871 (electronic)
Textile Science and Clothing Technology
ISBN 978-981-13-9580-2 ISBN 978-981-13-9578-9 (eBook)
https://doi.org/10.1007/978-981-13-9578-9

This Springer imprint is published by the registered company Springer Nature Singapore Pte Ltd.
The registered company address is: 152 Beach Road, #21-01/04 Gateway East, Singapore 189721,
Singapore

Contents

LCA (Life Cycle Assessment) on Recycled Polyester   1
Aravin Prince Periyasamy and Jiri Militky

Advancements in Recycled Polyesters 31
A. Saravanan and P. Senthil Kumar

Recycled Polyester—Tool for Savings in the Use of Virgin Raw
Material 49
Shanthi Radhakrishnan, Preethi Vetrivel, Aishwarya Vinodkumar
and Hareni Palanisamy

Case Studies on Recycled Polyesters and Different Applications 85
P. Senthil Kumar and P. R. Yaashikaa

LCA (Life Cycle Assessment) on Recycled Polyester

Aravin Prince Periyasamy and Jiri Militky

Abstract Polyester is a synthetic material which is produced from the petroleum products. The various environmental impacts are associated with polyester from manufacturing to end of life. Therefore, the manufacturing of recycled polyester (rPET) is an important to process as concerned with environmental impact and also inevitable. The rPET has a wide scope of their potential applications similar to virgin polyester. Generally, life cycle assessment (LCA) technique investigates the environmental impacts of the particular products from its cradle to grave. Therefore, it helps to identify the critical phase which creates the maximum impact on the entire product life cycle. So, it is significant to understand the environmental impact of rPET, nevertheless, LCA on rPET is foreseeable. The data from the LCA can initiate preliminary steps to reduce the environmental burdens from the products, also it provides the detailed information on how it affects the ecosystem. In this chapter we discussed about the LCA on rPET, initially, the brief introduction will be provided about the present manufacturing techniques of rPET. Various issues associated with sustainability of rPET manufacturing, importance and methodology of LCA on rPET were explained in detail. Based on the LCA results, the important parameters with respect to the sustainability of rPET would be present in this chapter.

Keywords Cradle to grave · Polyester · Recycling · LCA · GWP

1 Introduction

The several natural fibers were chiefly used for the production of textiles and garments until the seventeenth century [1, 2]. Either way today's situation differs, according to the development of synthetic fibers in the late 1930s, these fibers are now largely used for textiles [3]. Polyester fibers are the examples of synthetic fibers containing ester groups in their main polymeric chain [4]. Polyethylene terephthalate (PET) having the ester group and generally known as polyester. In 2018, 106 million tons of global

A. P. Periyasamy (✉) · J. Militky
Faculty of Textile Engineering, Department of Materials Engineering, Technical University of Liberec, Studentska 2, 46117 Liberec, Czech Republic
e-mail: aravin.prince@tul.cz

© Springer Nature Singapore Pte Ltd. 2020
S. S. Muthu (ed.), *Environmental Footprints of Recycled Polyester*, Textile Science and Clothing Technology, https://doi.org/10.1007/978-981-13-9578-9_1

production compared to 25% cotton fibers, however, polyethylene terephthalate produced and consumes higher than any other textile fibers [5]. Eco-friendly industries and eco-friendly industrial practice has been promoted by the awareness created on environmental concerns. In the case of environmental benefits, the classical *3R* can be implemented in the rPET industry, also it must be promoted to make awareness to the consumers [6, 7]. As it is known recycling is not new with vast history [8]. Last two decades, the awareness of sustainability and waste management results protection towards to the environment by practicing the more and more recycling process. Humanity poses large problems mainly due to plastic and polymeric waste in which crude oil is the first non-renewable materials which is the raw materials to produce various thermoplastic materials including the textile fibers. Majority of the synthetic materials consists of larger molecular size and rigid structure resulting non-biodegradable and non-decomposable. Accounting into the problems above it is advisory to recycle plastics and polymers and recycling motivates to decrease or lower landfill expenses, compared to virgin plastics recycled polymers are cheaper and further energy can be recovered from the plastic through various process [9]. 60% of the global PET produced with high-molecular weight further utilized to produce the textile fibers and 30% of PET is utilized to produce the bottles and other articles [10]. According to the reference [11], 70–80% of crude oil is used to produce virgin polyester, among them only 30–40% were recycled. Therefore, it is necessary to take attention which increase the recycling percentage, resulting the reduction of the environmental burdens by landfilling as well as carbon emission. In order to reduce our carbon footprint, larger companies receive tons and tons of paper and plastic which is recyclable. However, some of the statistics says 91% of plastics are not recycled, apart from that people buy millions of plastic bottles (food, beverages, water etc.) per minute [12]. In 2015, approximately 20% of textiles were reused in Sweden [13], in 2018 it can be increased to 40% as per the Swedish Environmental Protection Agency (SEPA) [14] and is predicted to increase further 20% in 2020. Figure 1 shown the recycled PET bottles% in various forms in the USA. Public imagination has been gathered by the idea of using recycled PET materials including bottles, molded articles, textile fibers, buttons etc. The concept of recycling has become green option, since it reduces the energy requirement for the production of virgin PET also reduce the consumption of non-renewable resources.

Generally, the textile wastes can be classified into three types, which are [15]:

- Wastes from pre-consumer stage
- Wastes from post manufacturing
- Wastes from post-consumer stage

Pre-consumer stage waste is defined as waste generated during the production, for example in textile productions, short fibers in the spinning, yarns in both weaving and knitting and fabrics from garment cutting and many. Generally, these wastes can be reused and produce different products, for example the short fibers were used to produce the coarser (i.e. thicker) fabric. After manufacturing, products having the defects are classified into the wastes from post manufacturing stage, perhaps it can

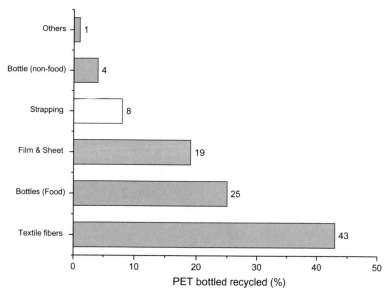

Fig. 1 Percentage of PET bottled recycled in various forms in the USA (2017)

be sold to very less price to the consumers or it will be reused and produced as same or different products.

To make rPET it includes the recycling of the accessories and beverage bottles as an example. Figure 2 summarizes the various routes of recycling and reuse of textiles. In any process waste is inadvertent and it has to be reused for the improvement of the environment. Recycling is the best solution for the polyester textile which drastically reduced the carbon emission and saves the energy as compared to virgin PET manufacturing.

If the materials were recovered from the waste use it again is called as reuse, whereas, after recovering, modify into the product is called recycling, meanwhile the recycled product is higher values than the original product is called as upcycling and lower values is called downcycling. There are four approaches for the recycling which is well explained in Fig. 3. Primary recycling defines the recycle of waste into original products. Recycling the post-consumer plastic waste into new products with reduced properties may classified into secondary recycling. Production of fuel or monomer from the PET waste are classified into tertiary recycling approach [17].

2   The Life Cycle Assessment Methodology

LCA is the method to evaluate the environmental performance of the products throughout the cycle, starting from raw material to it last stage of cycle. In accordance to the ISO standard, life cycle assessment can be conducted with four phases namely (Fig. 4) [18–20]. First stage comprises of aim of study and describe all the products that are assessed. In the second stage of inventory the raw material is acquired to the process of development to its final information of the product are collected [21]. In order to develop, calculate the data of discharge from the process of the life cycle of the product in this stage energy consumption, raw material requirement, environmental emission and discharged are figured and calculated. The third stage is impact assessment stage the data of inventory are translated into the effect of human health, ecological health and resource depletion. The last stage of the life cycle assessment is the interpretation where the results are interpreted and discussed. The four stages of LCA has been described graphically in Fig. 4.

- Goal and scope,
- Inventory analysis,
- Impact assessment and
- Interpretation.

The first phase is "goal and scope" where the purpose of LCA study can be well explained. It is clearly discussed in ISO 14040 and 14044 standards. Inventory analysis deals with the energy and material requirement to produce the products

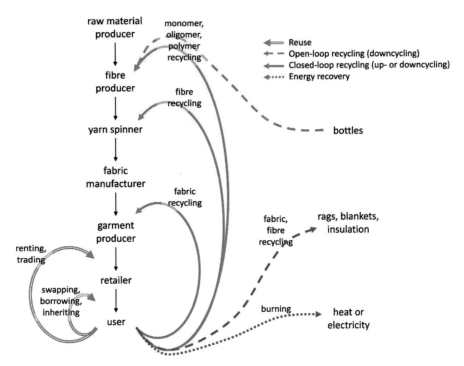

Fig. 2 Recycling and reuse routes for textile wastes, (reused under the terms of the Creative Commons Attribution license from Journal of Cleaner Productions, Elsevier Publications [16])

and its analysis from cradle to grave. Impact assessment can assess the significant potential effect on the environment with respect to throughout their life cycle of the product. Interpretation defines the discussion of the results.

Goal and Scope

In the life cycle assessment, goal and scope could be the first phase. If the goal and scope will be perfectly planned, the entire study could be easy. While planning for goal and scope, there are many things to be strictly considered, they are given below.

- The product system
- Function of the product system
- Functional units (rPET, virgin PET)
- Product system boundary
- Various LCA procedures
- Limitations of product system
- Critical review
- Requirement of data for this phase
- Hypothetical assumption of the product system

The scope of the study may require modification according to the product design and other parameters since LCA is an iterative technique further information is collected for every LCA study.

Life Cycle Inventory Analysis (LCI)

The LCA which is the second phase used in the data collection portion. It generally involves to quantifying the values of input and output products which includes energy, raw material and finished product. In the inventory analysis the following things must be in the check list.

- Raw materials
- Energy requirement
- Water
- Recourses (renewable or non-renewable)

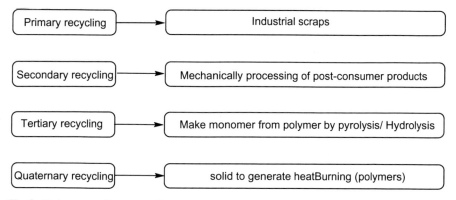

Fig. 3 Various recycling approaches

Fig. 4 The life cycle
assessment framework

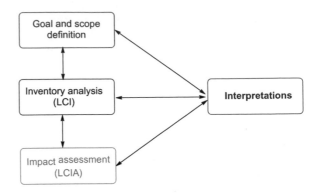

- Transportation
- Carbon and other substance emission
- Consumer use
- Disposal
- Landfilling
- Recycle or Reuse.

In case of decentralized sectors, it is challenging job to analyze, since every operation should not be in same place. For example, crude oil is available in the gulf countries, it transported into India to refine, thereafter virgin polyester is produced by using refined crude oil products. Then the produced fibers sent for the yarn and fabric formation. Later, garment manufacturing can be done in India or in Bangladesh, finally the garments were export into Europe or USA. If you consider in the supply chain, there are lot of transportation taking place which directly increase the carbon emission and energy utilization.

Impact Assessment (LCIA)
As discussed previous section, the impact analysis can quantify the potential impact on the environment due to the product formation. Generally, the results of the inputs and outputs are focused by this phase and categorize them strategically in order to help the environment. As per the ISO, this step can be done or follow with proper systematic procedure, the tentative procedure is given bellow, perhaps some of the steps are compulsory and some are to keep optional.

- Classification (compulsory)
- Characterization (compulsory)
- Normalization (optional)
- Grouping (optional)
- Weighting (optional)

Figure 5, explains the feasible implementation of a combined midpoint and damage approach which including the ecosystem quality, human health and climate changes [22].

Interpretation

LCA's last phase is the life cycle interpretation in which "identify, quantify, check and evaluate", previous phases such as life cycle inventory (LCI) and or the life cycle impact assessment (LCIA) [23] are the results from the information above. The Interpretation is made throughout all the phases with the purpose to summarize and discuss the results achieved systematically and to verify if the results are in accordance with the defined goal and scope. Final conclusion can be made in this phase after considering the following points.

- Identification of critical places
- Identification of potential significance on the environmental impacts.
- Determination of data sensitivity from the previous phases (LCI and LCIA).
- Conclusion
- Recommendation for the further analysis

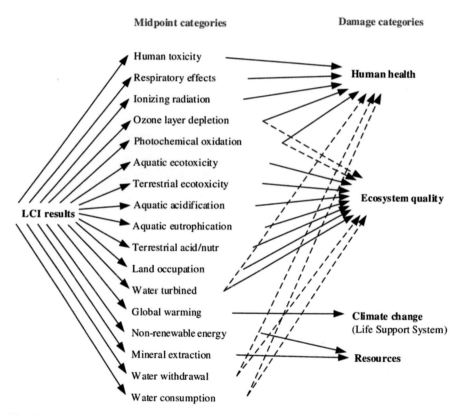

Fig. 5 Feasible implementation of a combined midpoint and damage approach, (reused under the terms of the Creative Commons Attribution license from Environmental and Climate Technologies, The Journal of Riga Technical University [22])

3 Product Life Cycle

The main purpose of the life cycle assessment is to reduce the utilization of various resources and emission of GHG and other substance throughout its life cycle [24]. Figure 8 describes the entire life cycle of rPET. Generally, the polyester waste can be recycled in two methods such as, 'open and closed loop' [25]. Simple explanation for the closed-loop method of recycling, if the product can be recycled to back (i.e. same product) is called open-loop recycling (e.g. bottle to bottle). If it recycled into different product is called as closed loop (e.g. bottle to fiber) [26–28]. Figures 6 and 7 exhibits the open-loop and closed-loop methods of recycling of PET respectively.

System Boundary
A system boundary is the important part in the life cycle assessment, it explains what are the things going to include in the LCA study [9, 29]. In case of rPET, the system boundary includes from the collection of wastes from various places (e.g. municipal centers) and it transport into the manufacturing hubs (i.e. melt spinning or melt-blown units) [28, 30, 31].

3.1 Raw Material Collection

Municipal waste is the prime source for the recycling of PET, which contains the various waste materials such as paperboards, brand labels, caps, lids, glass, plastic materials and metals. However, some of the city and residential hubs have the plan to keep these plastic containers in separately, which is not mixed with these wastes, it makes easy to separate and sort. Recovering plastic bottle packaging from residences has been reliable waste collection strategy, however it recovers only 20–45% [32–35]. Countries like India, the main sources for the collection of these plastic bottles are cafes, airports, railway stations and bus depot [30, 34, 36].

Sorting
Sorting is the important process in the recycling sequence, it is generally carried out by two method, either automatically or manually. While these processes, the unwanted (i.e. non-PET materials) materials can be sort outed. In terms of textile fabrics, polyester alone should be sorted, also ensures that the sewing thread must be in the same as polyester if not it should be separated. In the case of food and beverage PET bottles, the colored bottles must be separated from colorless one. Most of these bottles have the polyolefin caps (lids), which is necessary to separate from the batch. Usually the color sorting is very easy process [37, 38]. In automatic sorting machines consists of color sensible sensor which can be used for sorting the colored and colorless materials [39–41]. Fourier-transform near-infrared spectroscopy for polymer structure study can help to evaluate the mixture of non-PET materials [37, 38, 42].

Cleaning
After sorting the recycling materials were sent to cleaning process. It is necessary to clean the residual food items, beverages, short fibers, labels and other materials like

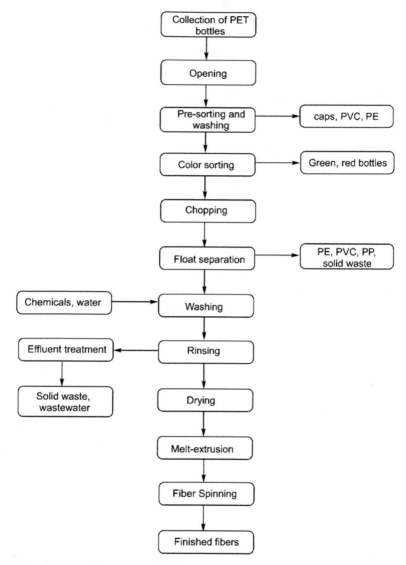

Fig. 6 Open-loop recycling process of PET

adhesives [43–46]. Generally, the municipal waste requires more water for cleaning, since it also calculated in the water footprint.

Final Separation
The final separation is the process where the PET can be separated from sink or floating materials such as threads, labels, adhesives from the PET flakes [47–49]. Placing the Raman emission spectroscopic detectors reduce the PVC contamination in the PET flakes [50, 51]. For PET flake sorting involves many methods which

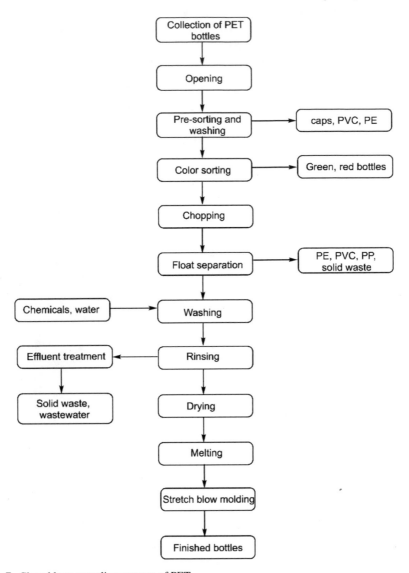

Fig. 7 Closed loop recycling process of PET

includes the conventional and modern methods, however there are some limitation in terms of conventional methods such improper colored flake sorting and difficult to sort the flake with different physical properties (i.e. variable density). The best example for modern flake sorting is laser-sorting [50, 52, 53]. Laser-sorting involves to sort the flakes by using of emission spectroscopy. Generally, the automotive plastic materials consist of different physical properties, but in this method can be sort easily. Even it can separate the same density material such as polyethylene from polypropylene [52–56].

3.2 Melt-Spinning

Recycling of PET can be carried out according to different strategies it has been discussed deeply in the previous section of this chapter. Due to the molecular weight (MW) reduction, the chemical recycling for the depolymerization of post-consumed PET to make monomers which can be used as raw materials, generally these can be reprocessed after several washing and few mechanical recycling process (i.e. grinding) [57]. During the chemical recycling, the PET undergoes different chain reaction due to thermal, mechanical and hydrolytic process.

3.3 Textile Manufacturing

Textile manufacturing is the series of process, here the fiber is fed as raw materials which converting into yarn by using of various spinning machineries. Then the yarn is fed as raw materials to produce the fabric (i.e. knitted or woven) and this process is called weaving or knitting. Later the fabric is converted into garments. Figure 8 describes the whole life cycle of rPET garments from cradle to grave.

Spinning for Yarn Formation
Spinning is the process which involves to produce the yarn from the fibers. Currently there are several spinning methods involved to produce the yarn. Due to the greater number of machineries (i.e. blowroom, carding, draw frame, comber preparation, comber, speed frame and ring frame), this process consumes high level energy. In spinning there are possibility for the flying of micro and macro fibers, which generally cause the byssinosis [6, 7, 21, 58–63].

Fabric Formation
The fabric can be prepared via weaving or knitting. The weaving is a process of formation of fabric with interlacement of two or more sets of yarns using a stable machine called loom. Human beings have started using the woven fabrics since the dawn of history. For weaving, loom or weaving machine is used to the fabric construction. Weaving process can be carried out after warping and sizing. In the weaving, there are many fabric weave patterns (i.e. designs) available, however, the plain weave is the simplest way of fabric construction in fabric weaving. In this weave, each weft yarns passes under and over the warp yarn across the width of the fabric. Similar to spinning, weaving also consumes high level of energy. Knitting is the technique of forming the fabrics, which has achieved tremendous success all over the world. A woven fabric has two sets of threads which cross each other at right

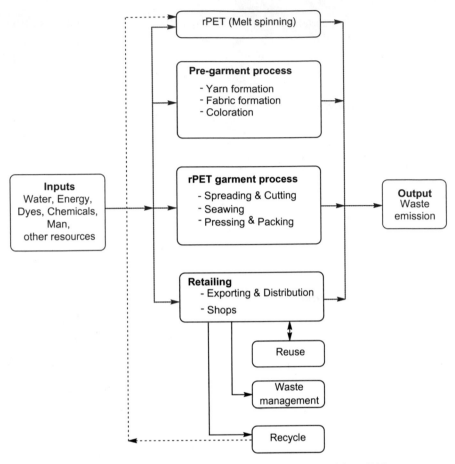

Fig. 8 Production life cycle for the garments made from rPET, (inspired from [21])

angles, producing a fabric that is relatively inelastic. A knitted fabric has been formed with only one thread or possibly a series of threads, and the fabric is composed of series of inter-looped loops. Since these loops can be distorted to a certain degree in any direction and since nearly all knitting yarns are more or less elastic, a knitted fabric is quite elastic.

Coloration
The term 'coloration' means 'to add color'. Textiles are usually colored to make them attractive for aesthetic appeal. They are also colored for functional reasons, e.g. military camouflage and fluorescent jackets for traffic or police personnel. Life would be more hazardous and certainly dull if textiles were used only in their natural colors. During the coloration of the recycled fabric, the dyes, water and chemicals are used which is the major environmental impacts in the coloration process. Prior to coloration, it is required to remove the impurities present in the fabric, therefore scouring of rPET fabric can be carried out. In the scouring, strong or mild alkalis

were used with warmer condition [6, 7, 21, 58–63], Perhaps, rPET or virgin PET, both are hydrophobic fibers as well as no presence of polar groups, resulting poor dye ability, therefore, it required a certain temperature (i.e. above 120 °C) for better dyeability. Nevertheless, it required lots of energy as well as increasing the GHG emission due to energy production. As environmental and health concerns, these dyes create waste water and release the harmful substances into nearby water bodies which is highly toxic [6, 7, 21, 58–63].

Garmenting
"Garment manufacturing" is the final process in the textile process sequence. In this garment manufacturing, there are numerous operations involved namely spreading, cutting, sewing, fusing, ironing and packing, also this process are more labor oriented process as compared to yarn and fabric manufacturing. In Asian region, the garment factories are located far away from spinning and weaving department, on other hand, Europe and USA were importing the garments from in these regions. For both raw fabric and finished garment requires to transport huge distance which results in higher energy consumption and followed by increasing the CO_2 emission. Around 8% is estimated to be the CO_2 emissions from freight transport. Eventually, transport costs increase due to the price of oil and gas where transportation to the huge consumer place which directly influenced the energy utilization. Recently the "green logistics" become popular due to the reduction of GWP and CO_2 by reduction of the distance between manufacturing hub to consumer place [7, 21, 64–67].

Consumer Use and Disposal
Consumer use and disposal phase is one of the important phases in the life cycle assessment. Since we all think that the majority of GWP and carbon emission from the manufacturing, but consumer disposal phase generating the almost equaling amount of the carbon emission like the production phase. Every year, the purchasing capacity of the people has been increased or sometimes they provide some better offer initiate to buy more and more garments per person. After consumer use, discarded garments may throw into municipal waste and finally it goes for the landfilling. Landfilling has been increased every year, however some of the organization makes awareness to the consumers to donate their clothes, which turns into partial success. Due to donating or reusing makes sense which reduce the landfill burdens. But for synthetic materials like polyester, there are huge possibilities to recycle into the same or different products. But usually the recycling ensures the higher or lower valued products. In this case, consumer attitude and behavior have to change for reduction of environmental impacts, the best was to reduce the buying of new products and make sure to reuse or recycle the same products [6, 59–61, 68]. In the concerns with environment, Levi's started the concept of "8 bottle one Jean", where 8 PET bottles recycled to produce the new Jeans [69].

Impact Categories (IC)
Impact category is one of the important concerns in the life cycle assessment, since it characterizes the environmental impacts. It can be selected in every life cycle assessment study to define the various environmental impacts caused by the rPET or

other products [70–73]. Currently the following six impact category is very important, they are,

- *Global warming potential (GWP)*, it also described as 'climatic change'. It is described in amount of CO_2 emission (kg CO_2 eq).
- *Ozone depletion potential (ODP)* is related to the decomposition of ozone layer due to environmental impacts. It can be measured through [kg CFC-11 eq].
- *Eutrophication Potential (EP)* is related to the quantification of available of nutrients present in the waterbodies (i.e. phosphorous and nitrogen). Sometime it is called as phosphorous footprint.
- *Acidification potential (AP)* is the releasing of protons (H^+ eq) in the ecosystems.
- *Photochemical ozone formation* is related to the degradation of volatile organic compounds (VOC). It causes various health and environment issues.
- *Primary energy consumption (PEI)* is related to the non-renewable and renewable energy utilization.

Apart from these important impact categories, there are several categories including,

- *Human toxicity*, defines the impact of human health due to the hazardous chemical utilization.
- *Aquatic toxicity*, related to the enrichment of aquatic ecosystems with nutrients with respect to the water quality.
- *Particulate matter.*
- *Noise & Light pollution.*

When we performing the LCA, the respective study might require more specific impact categories. Since, if we need to achieve the proper results might require the relevant and respective impact categories can fulfil.

4   Life Cycle Assessment of RPET

LCA on the Open-Loop Recycled PET

Shen et al. [29] studied the LCA of open-loop recycled PET. In this recycling, cradle stage is very difficult to define, therefore they choose the cut-off approach, which could easy to define the system boundary. The detailed system boundary is described in Fig. 9. Here life can be split as first and second life which is purely independent.

From their results [29], Table 1 shows LCA results of rPET from cradle to gate based on the cut-off approach. As compared to virgin PET, rPET fibers save around 45–85% of non-renewable energy utilization. Resulting, rPET has a strong influence on the reduction of global warming potential (GWP). Results show that there is a reduction of 76% from mechanical recycling, 54% via semi-mechanical recycling, 36% and 24% via chemical recycling of PET with oligomer and monomer forms respectively. Therefore, it concludes that these recycling process reduced

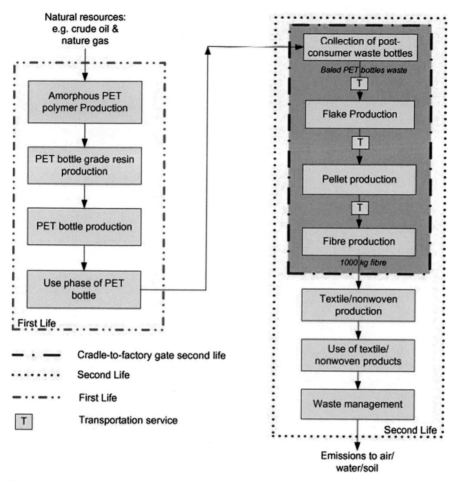

Fig. 9 Cradle to grave of open-loop recycled PET with cut-off approach, (reused from [29], with kind permission of Elsevier Publications)

significantly the environmental impacts with respect to all seven Centrum voor Milieuwetenschappen Leiden (CML) categories which excluding the freshwater aquatic ecotoxicity. As compared to another recycling process, the mechanical type of recycling shows the lowest environmental impacts. Since for the production of fibers, need to increase the PET to above the melting temperature, where there is a higher energy utilization, resulting in the responsible environmental impacts, particularly, abiotic depletion, photooxidation and acidification etc. Meanwhile, there is a requirement of water for cleaning the PET bottles and other process makes wastewater, which necessary to reduce the chemical oxygen demand. The wastewater is responsible for the eutrophication followed the toxicity due to the mixing of untreated water with nearby areas. Allocation factor (AF) can be calculated through waste evaluation technique, which helps further to identify the baled PET waste and

Table 1 LCA result of open-loop recycled PET fiber, based on the "cut-off" approach (reused from [29], with kind permission of Elsevier Publications)

Recycling route	M	S-M	Ch-BHET	Ch-DMT	V-PET
Company	Wellman	LJG	FENC	N/A	N/A
Fiber type	Staple	POY	POY	POY	Staple or POY
Non-renewable energy use (GJ equiv.)	13	23	39	51	95
Global warming potential 100a (t CO_2 equiv.)	0.96	1.88	2.59	3.08	4.06
Abiotic depletion (kg Sb equiv.)	6	11	18	N/A	45
Acidification (kg SO_2 equiv.)	3	9	14		21
Eutrophication (kg PO_4^{3-} equiv.)	0.8	0.7	2.3		1.2
Human toxicity (kg 1,4-DB equiv.)	362	415	745		4393
Fresh water aquatic ecotoxicity (kg 1,4-DB equiv.)	296	250	303		58
Terrestrial ecotoxicity (kg 1,4-B equiv.)	7	7	17		12
Photochemical oxidant formation (kg C_2H_4 equiv.)	0.2	0.3	0.6		1.0

M mechanical recycling; *S-M* semi-mechanical recycling; *Ch-BHET* chemical recycling with the monomer (BHET); *Ch-DMT* chemical recycling in the monomer (DMT) and *V-PET* virgin PET fiber

virgin PET materials. The AF is the ratio of market value of waste PET bottle waste and the market value of virgin PET bottle grade resin. However, the AF could have varied with respect to the continent, amount of waste generations and most importantly on the crude oil price. Due to the fluctuation of crude oil price, sometimes it is difficult to calculate the accrete value, nevertheless, we can add some percentage in the tolerance. With respect to the AF, they compare the LCA results and it shown in Table 2.

Mechanical and semi-mechanical recycled PET fibers supplies benefit to the environment in all category than that of virgin fiber except in freshwater aquatic ecotoxicity. Analyzing back-to-monomer recycling based on the "waste valuation" method is not being possible due to lack of sufficient data. The rPET from bottle offers higher environmental benefits than the single-use virgin PET fiber. Among the various types of recycling, the mechanical rPET cannot be recycled for further, perhaps, the chemical rPET can be further recycled even under the larger scale with economic viability.

Table 2 LCA result of open-loop recycled PET fiber, based on the "waste valuation" approach, (reused from [29], with kind permission of Elsevier Publications)

Mechanical	S-M	Ch-BHET	V-PET fiber	
Company	Wellman	LJG	FENC	
Type of fiber	Staple	POY	POY	Staple/POY
NREU (GJ equiv.)	40	49	66	95
GWP100a (t CO_2 equiv.)	2.03	2.95	3.66	4.06
Abiotic depletion (kg Sb equiv.)	19	23	31	45
Acidification (kg SO_2 equiv.)	8	14	19	21
Eutrophication (kg PO_4^{3-} equiv.)	1.1	1	2.6	1.2
Human toxicity (kg 1,4-DB equiv.)	1640	1700	2030	4390
Fresh water aquatic ecotoxicity (kg 1,4-DB equiv.)	300	250	305	58
Terrestrial ecotoxicity (kg 1,4-DB equiv.)	8	7	17	12
Photochemical oxidant formation (kg C_2H_4 equiv.)	0.4	0.6	0.8	1

However, the bottle to bottle recycling is another way of recycling the PET bottles which is an example of closed-loop recycling system.

LCA on the Closed-Loop Recycled PET
Chilton et al. [30] conducted the LCA on PET bottles under closed-loop method. Figure 10 explains the system boundary as well as the closed-loop recycling process for the respective studies. Table 3 described the overall emission results of rPET. In each process steps, PET cleaning is the highest responsible for the GWP, since it emits the various GHG such as CO_2, CO, SO_2, and NO_x, particulate material and dioxins. This is due to the utilization of electrical energy, of course while production of the electrical energy emits the huge quantity of GHG. Apart from that, transportation of PET from material reclamation facility (MRF) to processing plant is another source for the generation of HCl, ammonia, lead and cadmium.

LCA on the Recycled PET Post-Consumer Use
Intini et al. [74] studied the thermal insulation panel from recycled PET bottles and fiber which is obtained from post-consumer use. In this case, they make a nonwoven fabric out of PET waste materials to use in the thermal insulation panel. The detailed system boundary is given in Fig. 11. Separate collection increases the percentage of recovered PET from nonwoven manufacturing, it would be reduced the amount of purchased waste PET from other countries which reduce the transport and followed by the GHG emissions. In place of acrylic resin, the use of thermo-bonding virgin PET provides the better viscosity of the product. While maintaining the higher thermo physical properties of recycled products, there will be drastic reduction of environmental impacts due to the virgin PET, which involved by the use of waste PET which is shown in the results Table 4 [74]. Table 4 discussed the environmental impact of rPET flakes, rPET fiber and rPET thermal insulated panel, which is

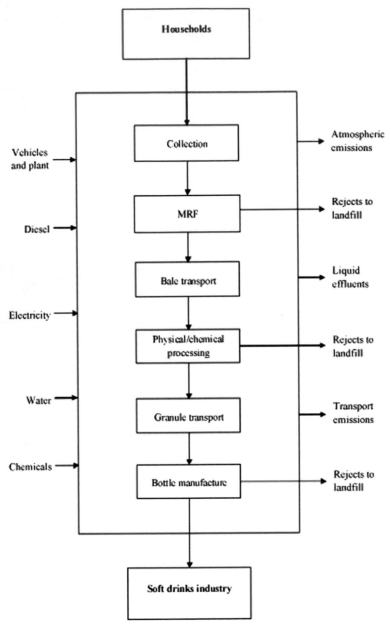

Fig. 10 The system boundary for closed-loop PET recycling, (reused from [30], with kind permission of Elsevier Publications)

Table 3 Emission from PET recycling, (reused from [30], with the kind permission of Elsevier Publications)

Pollutant	PET collection	MRF	PET cleaning	PET transport	Bottle manufacture
CO_2 (kg)	4	4	76	12	5
CO (kg)	0	7	85	0	8
NO_x (kg)	20	4	50	20	6
SO_2 (kg)	0	6	81	6	8
HCl (g)	6	1	33	60	1
Particulates (g)	0	5	66	22	7
VOC (g)	70	1	6	22	1
NH_3 (g)	5	0	0	94	0
Pb (mg)	1	0	0	99	0
Cd (mg)	6	1	2	90	0
Hg (mg)	1	0	81	17	0
Dioxins (μg)	11	5	63	14	6

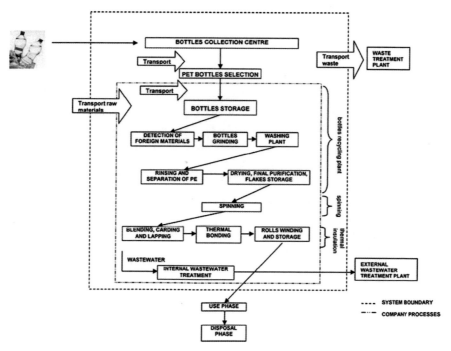

Fig. 11 The system boundary for PET recycling for post-consumer use, (reused from [74], with the kind permission of Springer Nature)

Table 4 Impact category results of PET recycling for post-consumer use (reused from [74], with the kind permission of Springer Nature)

Impact category	Unit	1 kg of recycled PET flakes	1 kg of recycle PET fiber	1 kg of thermal insulation panel
Global warming potential (GWP100)	kg CO_2 eq	0.365	0.901	1.675
Ozone layer depletion (ODP)	mg CFC-11 eq	0.000	0.000	0.000
Photochemical oxidation	g C_2H_4 eq	0.000	0.001	0.001
Acidification	kg SO_2 eq	0.001	0.002	0.005
Eutrophication	gPO_4 eq	0.000	0.001	0.002

assessed by the first analysis. From this, it is found that, there is a very low impact of rPET flakes as compared to the virgin PET. Perhaps, collection and transporting the waste PET bottle accounts a 6% of GWP for one kg PET flakes, whereas over 20% same impact category takes over the additives influence.

LCA on the Bio-based and Recycled PET

Shen et al. [75] associated the life cycle energy and GHG emission of bio-based PET, recycled PET, PLA and other man-made cellulosic fibers such as viscose rayon, modal rayon and Tencel. The typical system boundary is shown in Fig. 12.

Figures 13 and 14 show [75] the results of cradle-to-grave non-renewable energy use (NREU) and GHG emissions. Recycled PET uses lower non-renewable energy utilization than other fibers, even less than the Tencel fiber.

Figure 14 describes the cradle to grave of PET and other polymers on GHG. Recycled PET shows much lower GHG emission as compared to virgin Tencel and virgin PLA, of course, petrochemical PET and bio-based PET.

LCA on the Textile Manufacturing

Kang et al. [76] studied the LCA on the different process of textile manufacturing. Table 5 shows the inventory for LCA of polyester with respect to each production stage in both inputs (i.e. resources, water, products with their energy consumption) and output (i.e. wastewater, gas and solid waste) are established by the inventory analysis. The data for cotton cultivation, transportation of material from cultivation to the production factories and production factories to the consumer or distributors, textile spinning, pre-treatment, dyeing, printing is calculated from the previous researchers [77–79]. The environmental impacts of the recycling process are ignored as it was assumed that half of it is reused whereas another half is incinerated or used in sanitary landfill. In their study, they calculated the gross coal consumption by dividing the direct coal consumption for the energy production as well as the steam consumption for the production. Among the various process, the coloration of textile phase consumes huge energy as well as large requirement of resources (Table 6), on other hand it produce vast quantity of solid and liquid waste.

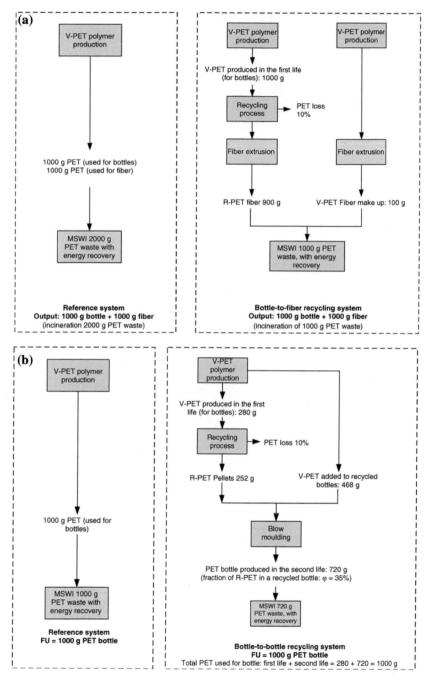

Fig. 12 The system expansion method to closed loop recycling (a) and open loop recycling (b). Where V-PET virgin PET; R-PET recycled PET, (reused from [75], with the kind permission of John Wiley and Sons)

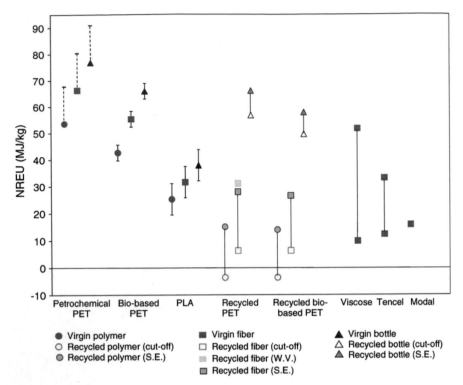

Fig. 13 Comparison of cradle to NREU of PET and other polymers, (reused from [75], with the kind permission of John Wiley and Sons)

Future Scope and Conclusion

The world's production of PET is accounted for 60% which is twice that of the plastic bottles which are developed from the non-virgin supply chain for polyester fiber. It has the greatest impact on the global energy and resource requirements, as a source of raw materials used recycled PET petroleum in the lesser amount. In a quality wise the rPET as same like virgin PET, yet the production of rPET requires 59% lesser energy than the virgin PET. Additionally, discards are being curbed which is turn prolongs landfill life and reduce toxic emission from incinerators. The environmental impacts of recycling bottle-to-fiber (open loop methods) are discussed in this chapter. The authors [29] used three techniques, such as cut-off, waste evaluation and system expansion. The system boundary of the cradle-to-factory gate was followed by the cut-off, system evaluation and the waste evaluation methods are analyzed. The non-renewable energy saving of 40–85% from the rPET fiber based on all three techniques as compared to virgin PET. Also, the rPET offers the reduction of GWP from 25 to 75% depends on the type of technology. In simple words, most of the environmental categories are reduced by the impact of bottle-to-fiber recycling based on all three techniques. Also, the authors analyzed the impact of rPET which is produced via different methods such as mechanical, semi-mechanical, chemical- monomer and

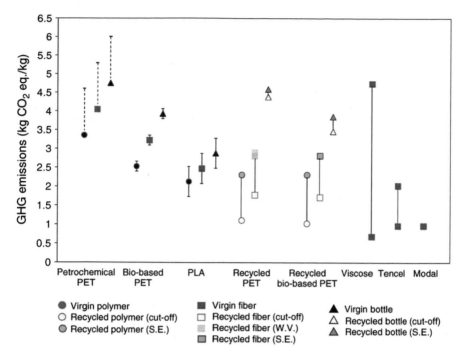

Fig. 14 Comparison of cradle to GHG of PET and other polymers, (reused from [75], with the kind permission of John Wiley and Sons)

chemical-oligomer. The rPET fiber has diverse applications when it produced from chemical recycling process than the semi-mechanical recycled PET. As concerns with environmental impacts, the mechanical and semi-mechanical has a lower impact than the chemical route of recycling. The chemical recycling offers the highest product quality, meanwhile, it has the highest impact on the environmental impacts than the mechanical and semi-mechanical route.

The overall conclusion, rPET offers the net reduction of emission of CO_2, carbon monoxide, acids and their gases, particulate matters, heavy metals and dioxins. This is due to the reduction of the landfill as well as emission associated with the manufacturing the virgin PET with respect to an equivalent amount. Of course, during the recycling process also involved the emission of above said gases and materials, among the various recycling process, the cleaning and transporting accounts huge share than the other process. Last but not least, the LCA studies can provide information which helps to make the decision process on the environmental impacts. However, the various other factors should be included while making the final conclusions, the factors are the cost of recycling, the market potential for recycled PET materials, national & international policy and regulations.

Regarding recycling, there are lots of facts which are important. Due to population increase, each of them adds a significant amount of waste in his life time to this

Table 5 LCA results of polyester cotton production (reused from [76], with the kind permission of Springer Nature)

Flow	Unit	Planting	Spinning and textile	Printing and dyeing	Final disposal
Inputs					
Pesticides	kg	98.7	0	0	0
Fertilizer	kg	34,387.5	0	0	0
Yarn	t	0	101	0	0
Water	t	10,000	24.2	13,010.6	0
Electricity	kWh	0	5626	95,087.9	0
Coal	t	0	10.2	17,732.6	0
Lye (NaOH)	t	0	0	14	0
Dyes	kg	0	0	16	0
Outputs					
Pesticides	kg	7	0	0	0
Fertilizer	kg	20,632.5	0	0	0
Dust	kg	0	2040	354,609	0
CO_2	kg	0	26724	4,645,310	421
SO_2	kg	0	86.7	151,143.8	0.6
NO_X	kg	0	75.5	13,122	0.4
CH_4	kg	0	0	0	13.1
Hg	kg	0	0	0	0.4
Wastewater	t	0	0	729.1	0
COD	kg	0	0	299	0
Lye (NaOH)	t	0	0	51,066	0
SS	kg	0	0	376.5	0

world. People produce enormous waste in a single day in New Delhi to fill up the whole Taj Mahal for two times. It is essential to create educational awareness to the customers. Many universities provide education of recycling as an option or a required course. In general, multidisciplinary engineering and science education in material science, chemistry, mechanical, chemical and environmental engineering are involved in this type of education. It has enormous job opportunities apart from environmental benefits. After waste are categorized according to paper, plastics, glass, etc., are further taken to recycling plants. The logistics and processing of waste are involved by man power which helps them in jobs addition to that of the welfare of their families.

For rRET the final step in value chain that is market and demand is a strong driver for increased collection. Virgin PET is substituted by the secondary PET to a high extent to low reach environmental impact of PET collection and recycling. Collection and recycling systems providing high-quality secondary PET raw material

Table 6 LCA inventory results of polyester cotton on various chemical processing (reused from [76], with kind permission of Springer Nature)

	Unit	A	B	C	D	E	F	G	H	I
Inputs										
Water	t	0	1207	24	0	23.8	0	44.7	0	1.1
Electricity	kWh	16.7	6896.6	208	862	904.8	0	175.4	444.4	0
Coal	t	0	1724	20.8	0.9	7	0.37	15.8	0.5	3.3
Lye (NaOH)	t	0	10	4	0	0	0	0	0	0
Dyes	kg	0	0	0	0	66	0	0	0	0
Outputs										
Dust	kg	0	344,800	4200	180	1430	74	3160	95	670
CO_2	kg	300	4,517,000	54,600	2400	18,700	970	41,370	1240	8730
SO_2	kg	0	14,700	177	7.8	60	3	143	4	28
NO_X	kg	0	12,760	154	6.8	53	2.7	117	3.5	25
Wastewater	t	0	698.6	0	0	0	0	39.5	0	0
COD	kg	0	282.8	0	0	0	0	16.2	0	0
Lye (NaOH)	t	0	48,300	0	0	0	0	2766	0	0
SS	kg	0	34.5	0	0	0	0	2	0	0

A Singeing; *B* desizing; *C* mercerizing; *D* preliminary finish; *E* pad dyeing; *F* hot melt fixation; *G* reduction fixation; *H* heating setting; and *I* preshrinking to the market are required products of high quality made by recycling. Cooperation and communication in the value chain are called. For these criteria of rPET in public procurement and giving economic incentives/advantages to product manufacturers using rPET are included with policy instruments focusing on creating a pull by using rPET. In order to broaden and deepen the conventional LCA to more comprehensive LCA are initiated by many of the recent developments in the LCA of rPET. Scope of current LCA from environmental mainly impacts only covering all three dimensions of sustainability (people, planet and prosperity) with a broadened framework as that of this regard. Product level, sector level and economy level the scope. are predominantly broadened by it. Leaving the current LCA on rPET has to widened from technology level into physical, economic and behavioral relations [80]. Current LCA of rPET can expand into new approaches of ISO 14040 series such as ISO 14067 (water footprinting), ISO 14045 (life cycle costing) and ISO 14025 (other types of environmental labels and claims).

Software for LCA

Obviously, there is a lot of software available on the market place. The most disseminated LCA software is the following;

- https://ghgprotocol.org/Third-Party-Databases/Boustead-Model
- http://www.gabi-software.com/ce-eu-english/index/
- https://simapro.com
- http://idematapp.com
- http://www.eiolca.net
- https://www.bloomberg.com/research/stocks/private/snapshot.asp?privcapId= 6771295
- https://www.genan.eu/about-us/life-cycle-assessment/
- https://www.ifu.com/en/umberto/lca-software/

References

1. Periyasamy AP, Dhurai B (2011) Salt free dying. Asian Dyer 8:47–50
2. Periyasamy AP, Mehta P (2013) Lyocell fibers for nonwovens. Chem Fibers Int 63
3. Sinclair R (2014) Understanding textile fibres and their properties: what is a textile fibre? In: Sinclair R (ed) Textiles and fashion: materials, design and technology. Woodhead Publishing, pp 3–27. https://doi.org/10.1016/b978-1-84569-931-4.00001-5
4. Kajiwara K, Ohta Y (2009) Synthetic textile fibers: structure, characteristics and identification. In: Houck MM (ed) Identification of textile fibers. Woodhead Publishing Series in Textiles. Elsevier, pp 68–87. https://doi.org/10.1533/9781845695651.1.68
5. Lenzing AG (2018) Innovative by nature
6. Periyasamy AP, Venkatesan H (2018) Eco-materials in textile finishing. In: Martínez LMT, Kharissova OV, Kharisov BI (eds) Handbook of ecomaterials. Springer International Publishing, Cham, pp 1–22. https://doi.org/10.1007/978-3-319-48281-1_55-1
7. Venkatesan H, Periyasamy AP (2017) Eco-fibers in the textile industry. In: Martínez LMT, Kharissova OV, Kharisov BI (eds) Handbook of ecomaterials. Springer International Publishing, Cham, pp 1–21. https://doi.org/10.1007/978-3-319-48281-1_25-1

8. Fletcher BL, Mackay ME (1996) A model of plastics recycling: does recycling reduce the amount of waste? Resour Conserv Recycl 17:141–151. https://doi.org/10.1016/0921-3449(96)01068-3

9. Song HS, Moon KS, Hyun JC (1999) A life-cycle assessment (LCA) study on the various recycle routes of pet bottles. Korean J Chem Eng 16:202–207. https://doi.org/10.1007/BF02706837

10. Sarioğlu E, Kaynak HK (2018) Ch. 2: PET bottle recycling for sustainable textiles. In: Kaynak HK, Camlibel NO (eds) Polyester—production, characterization and innovative applications. IntechOpen, Rijeka. https://doi.org/10.5772/intechopen.72589

11. Why is recycled polyester considered a sustainable textile? 2009

12. A million bottles a minute: world's plastic binge "as dangerous as climate change." (2017) The Guardian

13. Strand J (2015) Environmental impact of the Swedish textile consumption: a general LCA study. Swedish Environmental Research Institute

14. Textile and recycling (2019). http://www.swedishepa.se/Global-links/Search/?query=textiles. Accessed Feb 10

15. Sustainable textile material (2017)

16. Sandin G, Peters GM (2018) Environmental impact of textile reuse and recycling—a review. J Clean Prod 184:353–365. https://doi.org/10.1016/j.jclepro.2018.02.266

17. Leonas KK (2016) In: Muthu SS (ed) The use of recycled fibers in fashion and home products. Springer Singapore, Singapore, pp 55–77. https://doi.org/10.1007/978-981-10-2146-6_2

18. Beck A, Scheringer M, Hungerbühler K (2008) Fate modelling within LCA. Int J Life Cycle Assess 5:335–344. https://doi.org/10.1007/bf02978667

19. Bjørn A, Hauschild MZ (2017) Cradle to cradle and LCA. In: Hauschild MZ, Rosenbaum RK, Olsen SI (eds) Life cycle assessment: theory and practice. Springer International Publishing, Cham, pp 605–631. https://doi.org/10.1007/978-3-319-56475-3_25

20. Muthu SS (2016) Evaluation of sustainability in textile industry. In: Muthu SS (ed) Sustainability in the textile industry. Springer Singapore, Singapore, pp 9–15. https://doi.org/10.1007/978-981-10-2639-3_2

21. Periyasamy AP, Wiener J, Militky J (2017) Life-cycle assessment of denim. In: Muthu (SS) Sustainability in denim. Woodhead Publishing Limited, U.K., pp 83–110 https://doi.org/10.1016/b978-0-08-102043-2.00004-6

22. Kittipongvises S (2017) Assessment of environmental impacts of limestone quarrying operations in Thailand. Environ Clim Technol 20:67–83. https://doi.org/10.1515/rtuect-2017-0011

23. Skone TJ (2000) What is life cycle interpretation? In: Environmental progress, vol 19. American Institute of Chemical Engineers, pp 92–100. https://doi.org/10.1002/ep.670190207

24. Nakatani J, Fujii M, Moriguchi Y, Hirao M (2010) Life-cycle assessment of domestic and transboundary recycling of post-consumer PET bottles. Int J Life Cycle Assess 15:590–597. https://doi.org/10.1007/s11367-010-0189-y

25. Hackett T (2015) A comparative life cycle assessment of denim jeans and a cotton t-Shirt: the production of fast fashion essential items from cradle to gate. College of Agriculture at the University of Kentucky

26. Utracki LA (2011) Recycling and biodegradable blends. In: Utracki LA (ed) Commercial polymer blends. Springer US, Boston, MA, pp 469–484. https://doi.org/10.1007/978-1-4615-5789-0_22

27. Bonifazi G, Serranti S (2019) Recycling technologies. In: Meyers RA (ed) Encyclopedia of sustainability science and technology. Springer New York, New York, NY, pp 1–57. https://doi.org/10.1007/978-1-4939-2493-6_116-4

28. Gomes TS, Visconte LLY, Pacheco EBAV (2019) Life cycle assessment of polyethylene terephthalate packaging: an overview. J Polym Environ 27:533–548. https://doi.org/10.1007/s10924-019-01375-5

29. Shen L, Worrell E, Patel MK (2010) Open-loop recycling: a LCA case study of PET bottle-to-fibre recycling. Resour Conserv Recycl 55:34–52. https://doi.org/10.1016/j.resconrec.2010.06.014

30. Chilton T, Burnley S, Nesaratnam S (2010) A life cycle assessment of the closed-loop recycling and thermal recovery of post-consumer PET. Resour Conserv Recycl 54:1241–1249. https://doi.org/10.1016/j.resconrec.2010.04.002

31. Welle F (2011) Twenty years of PET bottle to bottle recycling—an overview. Resour Conserv Recycl 55:865–875. https://doi.org/10.1016/j.resconrec.2011.04.009

32. Woodard R, Bench M, Harder MK (2005) The development of a UK kerbside scheme using known practice. J Environ Manage 75:115–127. https://doi.org/10.1016/j.jenvman.2004.11.011

33. Cimpan C, Rothmann M, Hamelin L, Wenzel H (2015) Towards increased recycling of household waste: documenting cascading effects and material efficiency of commingled recyclables and biowaste collection. J Environ Manage 157:69–83. https://doi.org/10.1016/j.jenvman.2015.04.008

34. Dahlén L, Lagerkvist A (2010) Evaluation of recycling programmes in household waste collection systems. Waste Manage Res 28:577–586. https://doi.org/10.1177/0734242X09341193

35. Bach C, Dauchy X, Chagnon MC, Etienne S (2012) Chemical compounds and toxicological assessments of drinking water stored in polyethylene terephthalate (PET) bottles: a source of controversy reviewed. Water Res 46:571–583. https://doi.org/10.1016/j.watres.2011.11.062

36. Mansour AMH, Ali SA (2015) Reusing waste plastic bottles as an alternative sustainable building material. Energy Sustain Dev 24:79–85. https://doi.org/10.1016/j.esd.2014.11.001

37. Poon CS, Yu ATW, Ng LH (2001) On-site sorting of construction and demolition waste in Hong Kong. Resour Conserv Recycl 32:157–172. https://doi.org/10.1016/S0921-3449(01)00052-0

38. Dimitrakakis E, Janz A, Bilitewski B, Gidarakos E (2009) Small WEEE: determining recyclables and hazardous substances in plastics. J Hazard Mater 161:913–919. https://doi.org/10.1016/j.jhazmat.2008.04.054

39. Kuczenski B, Geyer R (2010) Material flow analysis of polyethylene terephthalate in the US, 1996-2007. Resour Conserv Recycl 54:1161–1169. https://doi.org/10.1016/j.resconrec.2010.03.013

40. Janajreh I, Alshrah M, Zamzam S (2015) Mechanical recycling of PVC plastic waste streams from cable industry: a case study. Sustain Cities Soc 18:13–20. https://doi.org/10.1016/j.scs.2015.05.003

41. Sadat-Shojai M, Bakhshandeh GR (2011) Recycling of PVC wastes. Polym Degrad Stab 96:404–415. https://doi.org/10.1016/j.polymdegradstab.2010.12.001

42. Hennebert P, Filella M (2018) WEEE plastic sorting for bromine essential to enforce EU regulation. Waste Manag 71:390–399. https://doi.org/10.1016/j.wasman.2017.09.031

43. Zhang M, Gao B (2013) Removal of arsenic, methylene blue, and phosphate by biochar/AlOOH nanocomposite. Chem Eng J 226:286–292. https://doi.org/10.1016/j.cej.2013.04.077

44. Khoo SC, Phang XY, Ng CM, Lim KL, Lam SS, Ma NL (2019) Recent technologies for treatment and recycling of used disposable baby diapers. Process Saf Environ Prot 123:116–129. https://doi.org/10.1016/j.psep.2018.12.016

45. Hole G, Hole AS (2019) Recycling as the way to greener production: a mini review. J Clean Prod 212:910–915. https://doi.org/10.1016/j.jclepro.2018.12.080

46. Barsky D, Sala R, Menéndez L, Toro-Moyano I (2015) Use and re-use: re-knapped flakes from the Mode 1 site of Fuente Nueva 3 (Orce, Andalucía, Spain). Quatern Int 361:21–33. https://doi.org/10.1016/j.quaint.2014.01.048

47. Wu G, Li J, Zhenming X (2013) Triboelectrostatic separation for granular plastic waste recycling: a review. Waste Manag 33:585–597. https://doi.org/10.1016/j.wasman.2012.10.014

48. Mueller W (2013) The effectiveness of recycling policy options: Waste diversion or just diversions? Waste Manag 33:508–518. https://doi.org/10.1016/j.wasman.2012.12.007

49. Bowman DJ, Bearman RA (2014) Coarse waste rejection through size based separation. Miner Eng 62:102–110. https://doi.org/10.1016/j.mineng.2013.12.018

50. Marques GA, Tenório JAS (2000) Use of froth flotation to separate PVC/PET mixtures. Waste Manag 20:265–269. https://doi.org/10.1016/S0956-053X(99)00333-5

51. Burat F, Güney A, Olgaç Kangal M (2009) Selective separation of virgin and post-consumer polymers (PET and PVC) by flotation method. Waste Manag 29:1807–1813. https://doi.org/10.1016/j.wasman.2008.12.018

52. Huang Y, Sutter E, Shi NN, Zheng J, Yang T, Englund D, Gao H-J, Sutter P (2015) Reliable exfoliation of large-area high-quality flakes of graphene and other two-dimensional materials. ACS Nano, vol 9. American Chemical Society, pp 10612–10620. https://doi.org/10.1021/acsnano.5b04258

53. Wehrl HF, Judenhofer MS, Wiehr S, Pichler BJ (2009) Pre-clinical PET/MR: technological advances and new perspectives in biomedical research. Eur J Nucl Med Mol Imaging 36:56–68. https://doi.org/10.1007/s00259-009-1078-0

54. Patra D, Mishra AK (2002) Recent developments in multi-component synchronous fluorescence scan analysis. TrAC Trends Anal Chem 21:787–798. https://doi.org/10.1016/S0165-9936(02)01201-3

55. Dias FB, Plomp L, Veldhuis JBJ (2000) Trends in polymer electrolytes for secondary lithium batteries. J Power Sour 88:169–191. https://doi.org/10.1016/S0378-7753(99)00529-7

56. Maris E, Aoussat A, Naffrechoux E, Froelich D (2012) Polymer tracer detection systems with UV fluorescence spectrometry to improve product recyclability. Miner Eng 29:77–88. https://doi.org/10.1016/j.mineng.2011.09.016

57. La M, Francesco P, Vinci M (1994) Recycling poly(ethyleneterephthalate). Polym Degrad Stab 45:121–125. https://doi.org/10.1016/0141-3910(94)90187-2

58. Periyasamy AP, Ramamoorthy SK, Rwawiire S, Zhao Y (2018) Sustainable wastewater treatment methods for textile industry. In: Muthu SS (ed) Sustainable innovations in apparel production. Springer Singapore, Singapore, pp 21–87. https://doi.org/10.1007/978-981-10-8591-8_2

59. Periyasamy AP, Rwahwire S, Zhao Y (2018) Environmental friendly textile processing. In: Martínez LMT, Kharissova OV, Kharisov BI (eds) Handbook of ecomaterials. Springer International Publishing, Cham, pp 1–38. https://doi.org/10.1007/978-3-319-48281-1_176-1

60. Periyasamy AP, Duraisamy G (2018) Carbon footprint on denim manufacturing. In: Martínez LMT, Kharissova OV, Kharisov BI (eds) Handbook of ecomaterials. Springer International Publishing, Cham, pp 1–18. https://doi.org/10.1007/978-3-319-48281-1_112-1

61. Periyasamy AP, Ramamoorthy SK, Lavate SS (2018) Eco-friendly denim processing. In: Martínez LMT, Kharissova OV, Kharisov BI (eds) Handbook of ecomaterials. Springer International Publishing, Cham, pp 1–21. https://doi.org/10.1007/978-3-319-48281-1_102-1

62. Periyasamy AP, Militky J (2017) Denim and consumers' phase of life cycle. In: Muthu SS (ed) Sustainability in denim. Woodhead Publishing Limited, U.K., pp 257–282. https://doi.org/10.1016/b978-0-08-102043-2.00010-1

63. Periyasamy AP, Militky J (2017) Denim processing and health hazards. In: Muthu SS (ed) Sustainability in denim. Woodhead Publishing Limited, U.K., pp 161–196. https://doi.org/10.1016/b978-0-08-102043-2.00007-1

64. Periyasamy AP (2018) Testing of chromic materials. In: Vikova M (ed) Chromic materials: fundamentals, measurements, and applications. Apple Academic Press, New Jersey, USA, pp 1–398

65. Ramamoorthy SK, Åkesson D, Rajan R, Periyasamy AP, Skrifvars M (2019) Mechanical performance of biofibers and their corresponding composites. In: Jawaid M, Thariq M, Saba N (eds) Mechanical and physical testing of biocomposites, fibre-reinforced composites and hybrid composites. Woodhead Publishing Series in Composites science and engineering. Elsevier, pp 259–292. https://doi.org/10.1016/b978-0-08-102292-4.00014-x

66. Rwahwire S, Tomkova B, Periyasamy AP, Kale BM (2019) Green thermoset reinforced biocomposites. In: Koronis G, Silva A (eds) Green composites for automotive applications. Woodhead Publishing Series in Composites science and engineering. Woodhead Publishing, pp 61–80. https://doi.org/10.1016/b978-0-08-102177-4.00003-3

67. Le CH, Louda P, Periyasamy A, Bakalova T, Kovacic V (2018) Flexural behavior of carbon textile-reinforced geopolymer composite thin plate. Fibers 6:87. https://doi.org/10.3390/fib6040087

68. Seipel S, Yu J, Periyasamy AP, Viková M, Vik M, Nierstrasz VA (2018) Inkjet printing and UV-LED curing of photochromic dyes for functional and smart textile applications. In: RSC advances, vol 8. The Royal Society of Chemistry, pp 28395–28404. https://doi.org/10.1039/c8ra05856c

69. Levi Strauss & Co. 2016. 8 bottles one jean. http://explore.levi.com/news/sustainability/introducing-levis-wasteless-8-bottles-1-jean/. Accessed Nov 15
70. Guinée JB (2015) Selection of impact categories and classification of LCI results to impact categories. In: Hauschild M, Huijbregts M (eds) Life cycle impact assessment. LCA compendium—the complete world of life cycle assessment. Springer, Dordrecht, pp 17–37. https://doi.org/10.1007/978-94-017-9744-3
71. Pfister S, Koehler A, Hellweg S (2009) Assessing the environmental impacts of freshwater consumption in LCA. In: Environmental science & technology, vol 43. American Chemical Society, pp 4098–4104. https://doi.org/10.1021/es802423e
72. McClelland SC, Arndt C, Gordon DR, Thoma G (2018) Type and number of environmental impact categories used in livestock life cycle assessment: a systematic review. Livest Sci 209:39–45. https://doi.org/10.1016/j.livsci.2018.01.008
73. Esnouf A, Heijungs R, Coste G, Latrille É, Steyer JP, Hélias A (2019) A tool to guide the selection of impact categories for LCA studies by using the representativeness index. Sci Total Environ 658:768–776. https://doi.org/10.1016/j.scitotenv.2018.12.194
74. Intini F, Kühtz S (2011) Recycling in buildings: an LCA case study of a thermal insulation panel made of polyester fiber, recycled from post-consumer PET bottles. Int J Life Cycle Assess 16:306–315. https://doi.org/10.1007/s11367-011-0267-9
75. Shen L, Worrell E, Patel MK (2012) Comparing life cycle energy and GHG emissions of bio-based PET, recycled PET, PLA, and man-made cellulosics. In: Biofuels, bioproducts and biorefining, vol 6. Wiley, New York, pp 625–639. https://doi.org/10.1002/bbb.1368
76. Kang H, Shao S, Zhang Y, Hou H, Sun X, Zhang S, Qin C (2018) Improved design for textile production process based on life cycle assessment. Clean Technol Environ Policy 20:1355–1365. https://doi.org/10.1007/s10098-018-1572-9
77. Baydar G, Ciliz N, Mammadov A (2015) Life cycle assessment of cotton textile products in Turkey. Resour Conserv Recycl 104:213–223. https://doi.org/10.1016/j.resconrec.2015.08.007
78. Rana S, Karunamoorthy S, Parveen S, Fangueiro R (2015) Life cycle assessment of cotton textiles and clothing. In: Muthu SS (ed) Handbook of life cycle assessment (LCA) of textiles and clothing. Woodhead Publishing, pp 195–216. https://doi.org/10.1016/b978-0-08-100169-1.00009-5
79. Li FG, Zhang LJ, Cui JJ, Dong HL, Zhang CJ, Wang GP (2005) Study of agricultural tri-dimension pollution on ecological system in cotton field and its control tactics. Cotton Sci 17:299–303
80. Guinée JB (2012) Life cycle assessment: past, present and future. In: International symposium on LCA and construction, pp 10–12

Advancements in Recycled Polyesters

A. Saravanan and P. Senthil Kumar

Abstract Plastics must be arranged for reusing, which includes exertion and cost. Research is centered on discovering substances that can encourage the blending of various sorts of plastics, known as compatibilizers, enabling them to be reused together. The greater part of plastics right now reused are made out of polyethylene terephthalate (PET), which is the segment utilized in most water bottles, and polyethylene, the most exceedingly created plastic. Chemical reusing strategies with lower vitality necessities, compatibilization of blended plastic wastes to stay away from the requirement for arranging, and growing reusing advancements to generally non-recyclable polymers. "New materials enter the market gradually, and consequently the greatest effect is in growing increasingly proficient techniques to reuse the plastics that are created in huge amounts today". The innovation of precisely arranging plastic waste has experienced a mechanical "upset", where old plastic is separated and used to make new items at a positive cost/advantage balance. Synthetic reusing is advancing quick with innovative developments for effective recouping vitality, creation of important new concoction items, for example, monomers or petrochemical feedstocks. Mechanical advances have been accomplished on the depolymerization of plastic waste to turn one sort of plastic into another that is progressively important. Warm and reactant breaking pyrolysis into fluid energizes is progressing with promising results. Inventive bioplastics which are completely recyclable and ecologically inviting are under extreme research in numerous modern and college research centers. Biorenewable parts for thermosets, supplanting hydrocarbon-based polymers with those produced using vegetable oils or other plant-based materials. That could prompt new end-of-life alternatives, for example, treating the soil or synthetic reusing for these materials.

A. Saravanan
Department of Biotechnology, Rajalakshmi Engineering College, Chennai 602105, India

P. Senthil Kumar (✉)
Department of Chemical Engineering, SSN College of Engineering, Chennai 603110, India
e-mail: senthilkumarp@ssn.edu.in

SSN-Centre for Radiation, Environmental Science and Technology (SSN-CREST), SSN College of Engineering, Chennai 603110, India

Keywords Recycle · Plastics · Chemical · Polyethylene terephthalate ·
Biorenewable

1 Recycled Polyester

In contrast to polyester, reused polyester utilizes PET as the crude material. This
is a similar material that is utilized in clear plastic water containers, and reusing it
to make the texture keeps it from going to landfill. The means associated with the
generation procedure are as per the following.

- The gathered PET jugs are cleaned, dried and pounded into little chips.
- The chips are warmed and went through a spinneret to shape strings of yarn.
- This yard is ended up in spools.
- The fibre is then gone through a creasing machine to make a fleecy wooly surface.
- This yarn is then baled, coloured and sewed into polyester texture.

Polyester is an artificial fibre, blended from petrochemical items by a procedure
called polymerization. With 49% of the worldwide fibre generation, polyester is the
most broadly utilized fibre in the attire area: in excess of 63,000 million tons of
polyester fiber are delivered every year. Creation of polyester texture includes huge
amounts of synthetic concoctions, crude materials and side effects that are harmful
and can dirty water and air and cause medical problems. It very well may be either
precisely or artificially reused, with feedstock comprising of either pre-or post-buyer
squander that can never again be utilized for its expected reason. This incorporates
returns of materials from the circulation chain [1].

Polyester is an extremely famous texture decision—it is, indeed, the most mainstream of the considerable number of synthetics. Since it can frequently have an
engineered feel, usually mixed with characteristic filaments, to get the advantage of
common strands, which inhale and feel great by the skin, combined with polyester's
toughness, water repellence and wrinkle obstruction [2]. Most sheets sold in the
United States, for example, are cotton/poly mixes. It is likewise utilized in the assembling of a wide range of attire and sportswear—also diapers, clean cushions, sleeping
pads, upholstery, drapes and rug. In the event that you take a gander at marks, you
may be shocked exactly what number of items throughout your life are produced
using polyester strands [3].

Fundamental polymer science isn't excessively entangled, however for a great
many people the production of the plastics that encompass us is a secret, which no
uncertainty suits the synthetic makers great. A working information of the standards
required here will make us progressively educated clients.

Polyester is just a single compound in a class of oil-inferred substances known
as polymers. In this manner, polyester (in the same way as most polymers) starts
its life presently as unrefined petroleum. Unrefined petroleum is a mixed drink of
segments that can be isolated by mechanical refining. Gas is one of these parts, and the
forerunners of polymers, for example, polyethylene are likewise present. Polymers
are made by synthetically responding a ton of little atoms together to make one long

particle, similar to a series of globules. The little atoms are called monomers and the long particles are called polymers.

The polymers themselves are hypothetically lifeless and subsequently not especially unsafe; however, this is definitely not valid for the monomers. Concoction organizations as a rule overplay how steady and inert the polymers are, yet that is not what we ought to be keen on. Polyester texture is delicate, smooth, and supple—yet still a plastic. It adds to our body trouble in manners that we are simply starting to get it. Furthermore, in light of the fact that polyester is very combustible, usually treated with a fire resistant, expanding the harmful burden. So in the event that you feel that you have experienced this long being presented to these synthetic substances and haven't had an issue, recall that the human body can just withstand so much harmful burden—and that the endocrine upsetting synthetics, which don't appear to trouble you might influence ages to come.

Again, this is a blog, which should cover themes in materials: polyester is by a long shot the most prevalent texture in the United States. Regardless of whether made of reused yarns, the poisonous monomers are yet the structure squares of the filaments. What's more, no notice is ever constructed of the preparing synthetic concoctions used to color and complete the polyester textures, which as we probably am aware contain a portion of the synthetic substances which are most harming to human wellbeing.

Recycled polyester is a generally new pattern in the eco style industry. It is a long way from supportable. Reused polyester is dangerous to the earth and the wearer. Among the pattern is reused plastic jugs being transformed into textures, which are viewed as practical to keep from them heaping in the landfills. On the off chance that plastic jugs do not have a place in the landfills, they positively do not have a place on the body.

The recycled plastic container fibre pattern is not sound. For one, not all plastic is BPA free. We at that point present our body to BPA. Through the extraction of BPA, Bisphenol is not expelled, and is similarly as lethal. Skin does not work like the liver, it does not channel poisons. The concoction poisons experience the skin into the circulation system and spread all through the framework. Engineered textures are hard to make tracks in an opposite direction from. We are being exposed to manufactured textures overwhelmingly from the abuse of the style business. Engineered texture is the most practical and productive to make, yet the unhealthiest for us to wear while it additionally dirties the earth.

When we buy reused plastic produced using plastic containers, we bolster its creation. The reused polyester piece of clothing is yet going to finish up in the landfill at any rate, taking more than 700 years to disintegrate. Another option in contrast to plastic is hemp plastic produced using biotechnology to supplant dangerous oil based plastic. The headway of reused manufactured strands through fibre innovation does not yield a propelled texture. As cognizant buyers, we should know about how fibre innovation might be a diversion from what the material is really produced using; an oil result.

An answer is wearing plant-based textures that surpass dead manufactured and dead creature filaments. Plants convey a vibration that help the living being. Plant-based filaments are the main hypoallergenic textures accessible. It has turned out to be progressively testing to try to achieve a 100% plant-based closet. Web based business and nearby green shops are strong about adding plant-based apparel to your closet. An exceedingly compelling answer for a sound way of life is the end of engineered filaments, which incorporate reused polyester. Wearing plant-based filaments safeguards and creates manageability of humankind, and the earth.

1.1 Favourable Circumstances to Reused Polyester

- Using more reused polyester decreases our reliance on oil as the crude material for our texture needs.
- Diverting PET jugs for this procedure diminishes landfill, and along these lines decreases soil sullying and air and water contamination, requires less vitality than virgin polyester.
- Garments made from reused polyester mean to be ceaselessly reused with no debasement of value, enabling us to limit wastage. This implies polyester article of clothing assembling could possibly turn into a shut circle framework.

1.2 Difficulties and Efforts

- Solving quality issues through pollutions of different added substances like cancer prevention agents, colors, stabilizers or hostile to blocking specialist and shortening of the polymer chain at de-polymerization organize.
- Finding substitutions for antimony, a polyester impetus known to be malignancy causing (conceivably 500 mg/kg PET).
- Ensuring consistent mechanical feedstock and shutting the circle by encouraging feedstock from material pre-and post buyer squander.
- Achieving recognizability and straightforwardness in the gathering, arranging and handling with social and reasonable conditions.
- Looking at lifecycle contemplations: biodegradability and recyclability of polymers.
- Transition towards inexhaustible biogenic feedstock transforming into mechanical filaments.

1.3 Limitations

Numerous pieces of clothing are not produced using polyester alone, yet rather a mix of polyester and different materials. Overall, it is progressively troublesome, if certainly feasible, to reuse them. "Now and again, it is actually conceivable, for instance mixes with polyester and cotton. In any case, it is still at the pilot level. The test is to discover forms that can be scaled up appropriately and we're not there yet", said Magruder to Suston Magazine a year ago. Certain overlays and finishings connected to the textures can likewise render them unrecyclable.

Indeed, even garments that are 100% polyester cannot be reused for eternity. There are two different ways to reuse PET: precisely and synthetically. "Mechanical reusing is taking a plastic jug, washing it, destroying it and afterward transforming it once more into a polyester chip, which at that point experiences the customary fiber making process. Concoction reusing is taking a waste plastic item and returning it to its unique monomers, which are indistinct from virgin polyester. Most rPET is acquired through mechanical reusing, as it is the less expensive of the two procedures and it requires no synthetic compounds other than the cleansers expected to clean the info materials. Notwithstanding, mechanical procedure, the fibre can lose its quality and hence should be blended with virgin fibre, says the Swiss Federal Office for the Environment.

1.4   Process of Recycling

The polyester chips produced by mechanical reusing can change in shading: some turn out fresh white, while others are rich yellow, making shading consistency hard to accomplish. "A few dyers think that its difficult to get a white, so they're utilizing chlorine-based dyes to brighten the base", she clarifies. "Irregularity of color take-up makes it difficult to get great cluster to-bunch shading consistency and this can prompt elevated amounts of re-colouring, which requires high water, vitality and synthetic use".

Moreover, a few examinations recommend that PET containers drain antimony, a substance "known to be disease causing", in the expressions of Textile Exchange on its site. Antimony oxide is regularly utilized as an impetus during the time spent making PET and polyester [4]. Wellbeing organizations around the globe state there is no explanation behind worry, as amounts are too little to be in any way viewed as harmful (500 mg/kg PET), yet even so Textile Exchange names "discovering substitutions for antimony" as one of rPET's "challenges". There is likewise a scholastic discussion concerning the computation of CO_2 emanations in the correlation between virgin polyester and rPET "in light of the fact that the effect of the fibre's first life is excluded in the in general ecological evaluation of reused strands. On the off chance that it would, results would contrast", as per the report from the Swiss Federal Office for the Environment.

1.5 Reused Polyester Discharges Microplastics

Last yet not least, some counter contend the assertion that rPET shields plastic from terminating in the seas. Regardless they complete a bit, as man-influenced textures to can discharge minute plastic filaments—the scandalous microplastics. As indicated by an ongoing report by a group from Plymouth University, in the UK, each cycle of a clothes washer could discharge in excess of 700,000 plastic filaments into the earth. A paper distributed in 2011 on the diary Environmental Science Technology found that microfibers made up 85% of human-made flotsam and jetsam on shorelines around the globe. It does not make a difference if pieces of clothing are from virgin or reused polyester, the two of them add to microplastics contamination [5].

2 Sustainability

Around 49% of the world's apparel is made of polyester, since the athleisure pattern has driven a developing number of customers to be keen on stretchier, progressively safe pieces of clothing. In any case, polyester is certifiably not a feasible material choice, as it is produced using polyethylene terephthalate (PET), the most well-known sort of plastic on the planet. So, most of our garments originate from unrefined petroleum, while the Intergovernmental Panel on Climate Change is calling for extreme activities to hold the world's temperature to a limit of 1.5 °C above pre-modern dimensions.

Recycled polyester, otherwise called rPET, is gotten by liquefying down existing plastic and returning it into new polyester fiber. While much consideration is given to rPET produced using plastic jugs and holders discarded by shoppers, as a general rule polyethylene terephthalate can be reused from both post-modern and post-purchaser input materials. Nevertheless, just to give a precedent, five soft drink bottles yield enough fibre for one additional expansive T-shirt. Despite the fact that reusing plastic sounds like an unquestionable smart thought, rPET's festival is a long way from being a unanimity in the economical style network. Fashion United has accumulated the principle contentions from the two sides [6].

Engineered strands are the most mainstream filaments on the planet—it is evaluated that synthetics represent about 65% of world creation versus 35% for characteristic filaments. Most engineered filaments are produced using polyester, and the polyester regularly utilized in materials is polyethylene terephthalate (PET). Utilized in a texture, it has regularly alluded to as "polyester" or "poly". Most of the world's PET creation—about 60%—is utilized to make strands for materials; about 30% is utilized to make bottles [7]. It is evaluated that it takes around 104 million barrels of oil for PET generation every year—that is 70 million barrels just to deliver the virgin polyester utilized in textures. That implies most polyester—70 million barrels worth—is made explicitly to be made into strands, NOT bottles, the same number of individuals think. Of the 30% of PET, which is utilized to make bottles, just a small portion is reused into strands. In any case, utilizing reused bottles—"occupying waste from landfills"—and transforming it into filaments has gotten the open's creative ability.

The reason-reused polyester (regularly composed rPET) is viewed as a green alternative in materials today is twofold, and the contention goes this way:

1. Vitality expected to make the rPET is not as much as what was expected to make the virgin polyester in any case, so we spare vitality.
2. We are keeping bottles and different plastics out of the landfills.

Sustainability is safeguarding of the environment of our earth. We are a piece of that eco-framework. We are not a no man's land. On the off chance that plastic does not have a place in a landfill, it should not have a place on our bodies. The reused polyester article of clothing is going to finish up in the landfills at any rate, and this makes an unsustainable item. There is right now an interest of plastic containers for the formation of reused polyester. Makers have even utilized new plastic jugs to create new fibre that is totally unimportant of it being reused. Because of the danger, plastic has by and large to our earth, wearing them isn't the arrangement.

Polyester fibre is ruling our style industry. Each segment of the style business: Outerwear, swimwear, active wear, formal wear, unmentionables and hosiery, and so forth. All are fundamentally adding to engineered material assembling. A considerable lot of them are presenting reused polyester from plastic containers. Numerous eco-accommodating brands are supporting the utilization of reused polyester, thinking about this item sheltered and reasonably designed. It is not reasonable. When we keep on assembling plastic jugs, we are approving the generation by re-making them as filaments. We are endeavouring to keep a material that takes more than 700 years to biodegrade.

Because of this blast of reused polyester, makers are making tech strands that endeavour to copy characteristic filaments. Indeed, even with fiber wicking innovation, these textures—regardless of how cutting-edge they show up—are not supporting your body. Poisons from texture medications and synthetic substances lock onto sweat and enter through the pores advancing into the circulatory system. Whenever synthetics and poisons get past the skin, our biggest organ, it sidesteps the liver. The liver helps channel and evacuate poisons. Whenever synthetic concoctions and poisons enter through the skin, there is no channel, which results in a more awful response and compound mien. In the event that we need to make a green planet (and be solid) a standout amongst the most imperative things we have to work with are plant strands. We should concentrate on constraining and notwithstanding dispensing with the creation of engineered dress. In the event that we need to make an amicable parity inside a maintainable earth and our framework we should wear normal filaments.

As society turns out to be increasingly acquainted with the risks related with sending old materials to the landfill, and as new reusing innovations create, it tends to be foreseen that the material reusing industry will keep on developing. In the meantime, watch for patterns, for example, moderate style to attract proceeded with thoughtfulness regarding the interchange of apparel and supportability. The quick design industry produces impressive contamination and a sizeable negative effect on environmental change. Buyers can help influence change by picking garments marks that last more and which exhibit a promise to diminishing their environmental change sway.

2.1 Environmental Issue

The ecological parts of the waste phase of apparel rely upon the strategy for transfer. Apparel is discarded in two different ways: with the household squander or through isolated accumulation. Separate accumulation prompts re-use, similar to second-hand or reusing as fabric, yarn, or even as fibre. The primary negative ecological effect of this is the discharge of CO_2 (ozone harming substance). A constructive outcome of consuming waste is the (conceivable) creation of vitality (steam, power, city-warming). Outside Western-Europe, garments still halfway winds up in landfills, prompting land use and potential emanations of dangerous substances to soil and ground water.

(i) **Reuse**

   Gathered articles of clothing are arranged and reused as items (about 58% in weight). About 20% in weight is prepared into clothes (for instance for cleaning). The reuse as an item or as clothes gives the material another (constrained) lifetime before it is finally tossed away. The fundamental advantage of reuse is that virgin crude materials for those items are spared.

   - Much of the still usable second hand garments is sold in second-hand shops or transported to creating nations and sold there on the neighbourhood showcase. An unwanted reaction can be that the nearby attire generation is undermined. A natural impact is that the garments end up at a low standard waste dump or cremation fire.
   - Sometimes new bits of dress are made from pieces of old garments. By joining and making new increases, the varied articles of clothing are promoted as a specific style.

(ii) **Recycling**

   The way toward reusing is as per the following:

   1. Approaching material is evaluated into sort and colour.
   2. The materials are destroyed and mixed with other chosen filaments, contingent upon the proposed utilization of the reused yarn.
   3. The mixed blend is cleaned and faded if important.

The utilization of reused material is particularly non-woven. It is utilized as sleeping cushion fillings, felt material for vehicle protection, material felts, amplifier cones, board linings and furniture cushioning. Turning of reused filaments is much more muddled than the non-woven application. The reusing procedure can be altogether unique for polyester and nylon. To make new polyester textures, new polyester fibre filaments are delivered by a procedure of destroying, grinding and softening pursued by expelling new tracks. The ecological advantage of reusing is that the fibre can essentially be reused repeatedly. This replaces the utilization of new virgin filaments with each cycle. In any case, the material fibre will systematically decline in quality and at last end up in waste transfer or burning (down cycling).

Reuse and reusing both give natural benefits:

- It decreases the requirement for landfill space.
- It decreases weight on virgin materials and non-inexhaustible assets (like raw petroleum).
- It for the most part results in less contamination and vitality use than generation from new crude materials.

One of the all the more encouraging improvements in economical materials is flax, a stalky and stringy plant that can be developed with far less water and less pesticides than cotton and created at a lower cost [8].

Reusing Process

For materials to be reused there are central contrasts among common and manufactured filaments. For common materials:

- The approaching unwearable material is arranged by kind of material and shading. Shading arranging results in a texture that should not be re-coloured [9]. The shading arranging implies no re-kicking the bucket is required, sparing vitality and maintaining a strategic distance from poisons. Materials are then maneuverer into strands or destroyed, in some cases bringing different filaments into the yarn. Materials are destroyed or maneuvered into filaments. Contingent upon the end utilization of the yarn, different filaments might be fused [10, 11].
- The yarn is then cleaned and blended through a checking procedure
- At that point the yarn is re-spun and prepared for ensuing use in weaving or sewing.
- A few strands are not spun into yards, in any case. Some are packed for material filling, for example, in sleeping pads.

Because of polyester-based materials, pieces of clothing are destroyed and after that granulated for preparing into polyester chips. These are consequently dissolved and used to make new strands for use in new polyester textures.

Two unique techniques can be utilized to make reused polyester. It tends to be made through a compound reusing course or it tends to be precisely delivered, which is the more typical course [12].

The Executives and Mechanical Techniques for Plastic Reusing

Overseeing and reusing of municipal solid waste (MSW) in numerous nations is protecting the earth as well as entirely gainful. Particularly extraction of different metals from the buildup (base fiery debris) following burning of MSW infers natural advantages as well as genuine financial focal points. Likewise, reusing of aluminum jars, paper and glass is mechanically feasible and beneficial. Conversely, plastic polymers reusing is frequently all the more testing in light of the fact that of low thickness and low estimation of waste. In each strategy for plastic reusing there are various specialized obstacles to survive. Plastic reusing is the way toward recuperating scrap or waste plastic and reprocessing the material into valuable new plastic items. Since most by far of plastic is non-biodegradable, reusing is a piece of worldwide endeavours to lessen plastic in the waste stream, particularly the roughly 8-10 million tons of waste plastic that enter the Earth's sea consistently. Plastic reusing incorporates taking any sort of plastic, arranging it into various polymers and after that chipping it into little drops or/and at that point softening it down into pellets.

Specific Arranging Strategies for Mechanical Plastic Reusing

Other arranging strategies that are utilized (or have been created in the last decade) for the isolating of blended polymers include:

Tribo-electric (Electrostatic) Division

It is appropriate for complex blends of plastic waste. The best outcomes have been accounted for division of a parallel blend like ABS/PC, PET/PVC and PP/PET ribo electrostatic partition is a viable procedure for the detachment of a blend of particles with moment contrasts in surface potential qualities and explicit gravity of its constituents. It is an electrostatic division, letting the polymer chips crash in a charging unit causes one to be charged decidedly and the other to be charged adversely. Numerous sorts of plastic have been isolated with high virtue by electrostatic detachment utilizing contact blender with rotational edges, fluidized-bed triboelectrification, or by utilizing vibrating feeder and electrostatic high voltage generator.

Foam Buoyancy or Particular Buoyancy Detachment

This foam buoyancy is another technique to isolate polymers with comparative densities. The essential standard of foam buoyancy is to have air bubbles follow (or not) to a chose polymer surface, hence making it skim.

Attractive Thickness Partition (MDS)

It is a refined thickness-based procedure, with its causes in the mineral preparing industry. By utilizing an attractive fluid (containing iron oxide) as the partition medium, the thickness of the fluid can be shifted by utilization of a unique attractive field. PET jugs are gathered to be reused. The PET is tainted with PP, PE, aluminum, glass, stones and different materials. In the MDS practice contaminants are evacuated in a succession of various partition units and the recouped PET can be reused in floor coverings, cushioning and garments. Be that as it may, there are still a few contaminants particles present in the PET. MDS can evacuate these particles, in this way creating a higher quality item while rearranging the procedure and decreasing procedure costs.

X-Beam Identification and Laser-Incited Plasma Spectroscopy

This is another helpful strategy for the partition of PVC holders, their high chlorine content makes them simple to recognize their sort. Spectroscopic partition can connect through a trademark backscattering from chlorine particles in PVC (x-beam fluorescent technique). Spectroscopic detachment additionally can be connected through investigation of the nuclear emanation lines produced by concentrating high-vitality laser radiations on plastics.

Depolymerisation of Plastic Waste

Mechanical advances have been accomplished in the most recent decade on the depolymerisation of plastic waste that even can make it conceivable to turn one kind of plastic into another that is progressively profitable. For example, in IBM's Almaden Research Focus analysts researched the mechanical parts of polycarbonates that can be depolymerise into their individual monomers with a base, for example, carbonate salts, and that a base likewise incorporated a high-esteem claim to fame plastic, polyether sulfone (PSU), utilized in restorative gadgets. The entire procedure should be possible in one stage with the equivalent base. Other research labs are exploring different avenues regarding the utilization of microorganisms to play out a reusing change of PET plastic waste. In this procedure PET can be pyrolized at high temperatures (450 °C) to deliver strong terephthalic corrosive (TA), and a microbiology group has found a few strains of microorganisms that can utilize TA as a wellspring of vitality and carbon to make a high-esteem biodegradable plastic called polyhydroxyalkanoate.

Thermal Cracking or Pyrolysis of Plastics

Thermal breaking or pyrolysis of plastics, includes the corruption of the polymeric materials by warming without oxygen (generally at temperatures 350–900 °C) bringing about the arrangement of a carbonized burn and a unpredictable division that might be isolated into condensable hydrocarbon oil comprising of paraffins, isoparaffins, olefins, naphthenes and aromatics, and a non-condensable high calorific esteem gas. The degree and the idea of these responses depend both on the response temperature. In any case, the warm debasement of polymers to low sub-atomic weight materials requires high temperatures and has a noteworthy downside in that an extremely wide item go is acquired. Synergist pyrolysis gives a way to address reusing issues.

Mechanical Recycling

Mechanical reusing is a technique by which squander materials are reused into "new" (auxiliary) crude materials without changing the fundamental structure of the material. It is otherwise called material reusing, material recuperation or, identified with plastics, back-to-plastics reusing. Post purchaser plastic waste can be an inhomogeneous and possibly defiled waste portion. It includes a tremendous scope of material sorts, with shape and size going generally, and as a rule the information material is made out of various material sorts (for example multilayer movies or composite things). The material passes broad manual or computerized mechanical arranging forms in specific offices, intended to isolate the extraordinary material streams. The best possible recognizable proof of materials is basic for accomplishing an amplified virtue of recyclates. In the wake of cleaning and granulating forms, the material is recouped by remelting and regranulating. The subsequent recyclates can be prepared with every single basic innovation of plastics change. In plastics reusing, the treatment of pre-purchaser (post-mechanical) material also, present customer material need on be recognized. On a fundamental level, the innovation of mechanical reusing is relevant to both bio-based traditional plastic and to most grades of biodegradable plastics. Specifically, the in-house reusing of mono-material pieces is rehearsed by and large both for regular plastics and for bioplastics.

Mechanical recycling utilizes clear plastic (polyethylene terephthalate) bottles that are likewise produced using PET pitch. The jugs are cleaned, chipped, softened and expelled into new PET filaments. The mechanical strategy utilizes post-buyer squander and less vitality amid preparing. PET jug reusing is more handy than numerous other plastic waste streams due to the high estimation of the plastic PET tar. Usually an unmistakable material that is moderately unadulterated and free from colorants and other practical added substances [13]. It must pass FDA prerequisites for nourishment bundling, which take out the capacity to include a great deal of risky synthetic concoctions. Plastic bottles made to convey water and other soda pops are made only from PET, which rearranges the arranging procedure at the reusing office.

Cooperation Between Bioplastics and Mechanical Reusing

In different nations, plastics accumulation and reusing frameworks are set up, focussing for the most part on the real plastics with high volumes, for example, PE, PET, PP and PS. Other than these, there is a wide scope of post buyer plastics, which are not being reused on account of their low volume or multifaceted nature (multi-layer films, mixes, composite things). In the event that a different reusing stream for a specific plastic/bioplastic type exists, the bioplastic can be effectively reused nearby its traditional partners (for example biobased PE in the PE-stream or biobased PET in the PET stream). The material facilitate arrangement of bioplastics beneath portrays normal bioplastics and how they are characterized by their biobased substance and biodegradability. While biobased reciprocals of traditional plastics can and do go into built up frameworks of mechanical reusing, different kinds of bioplastics can be incorporated relying upon the economy of scale. The waste administration frameworks should be improved considering the neighborhood foundation for gathering and reusing, nearby and territorial guidelines, the complete volume available and the organization of waste streams. There are general guidelines applying to ordinary just as to bioplastics to be considered:

(a) **Material Properties**

Most bioplastics can be made prepared for use in material reusing. In a few cases, contingent upon the conditions, extra advances are required. It may, for instance, be fundamental for PLA to experience an extra advance of polycondensation, or a unique crystallization arrange. What impacts the mechanical reusing procedure will have on the required properties of a certain polymer (or on given blends of polymers) under genuine conditions (for example post buyer material) is best replied by pragmatic experience.

(b) **Compatibility**

At whatever point various types of plastics items are reused (be it ordinary or biopolymers), the test of potential incongruencies between singular polymers must be considered. That is the reason a limit of consistency of the prepared material stream should be accomplished.

(c) **Economical Feasibility**

With the usage of close infrared examining strategies in the arranging of post shopper plastics, the discovery of bioplastic types and its sending to indicated material streams is conceivable and will prompt a dimension of virtue of recycles which is uncritical for most of utilizations. Mechanical reusing of plastics is to a lesser degree an issue of specialized plausibility yet more an issue of practical reasonability. The post shopper reusing of bioplastics for which no different stream yet exists, will be possible, when the business volumes and deals increment adequately to cover the speculations required. New independent streams (for example for PLA) will be acquainted in the short with medium term.

Advantages

- Material reusing of biobased plastics is effectively connected for postindustrial plastics (for example biobased PE and PET).
- Reusing of biobased plastics implies delayed carbon sequestration in items, therefore further improving their natural execution.
- The estimation of the polymer union as to vitality and other asset admission is protected.
- Mechanical reusing takes into consideration different lifecycles of a given plastic, subsequently substitutes and spares virgin material.

Challenges

The accompanying difficulties likewise apply to other new and little volume polymers entering the market.

- A market interest for the particular recycle is a fundamental precondition.
- Arranging and mechanical reusing of new polymers requires scale-up in volumes to be financially plausible and might require extra ventures.
- Mechanical reusing of post customer plastics squander requires suitable gathering, transport and arranging frameworks for clean also, homogenous waste streams. Such streams exist in numerous EU nations for PE or PET. Volumes for the bioplastics PLA a yet to develop to a specific dimension to make arranging (financially) feasible.

Chemical Recycling

Chemical Recycling includes utilizing synthetics to separate, or depolymerize, the polyester fibre back into its unique monomers, which would then be able to be polymerized over into new materials. The polyethylene terephthalate (PET) polymer is come back to its unique ethylene glycol and terephthalic corrosive monomers through synthetic responses. When it is polymerized once more into PET gum and polyester, the subsequent material is undefined from virgin polyester. This strategy is very costly and principally utilized for coloured and completed polyester items [14].

Plastics do have demonstrated advantages amid their utilization stage—for instance safeguarding of sustenance misfortune, lightweight development of vehicles and building protection. Plastic waste, nonetheless, and specifically plastic waste with regards to marine littering, is seen as a noteworthy worldwide test. There is additionally expanding administrative weight with respect to reusing amount and recyclability from one viewpoint and solid duties of our clients towards expanding the offer of reused material in their contributions then again. Understanding these difficulties requires development and joint endeavours all-inclusive over the esteem chain. With synthetic reusing, fossil assets for compound creation can be supplanted with reused material from plastic waste.

Plastic waste will be changed into a crude material utilizing thermochemical forms. The crude material can be sustained into the Verbund to make new concoction items with amazing item execution dependent on reused plastic waste. Substance reusing can help diminish the extent of plastic waste which winds up in landfill or burning. Be that as it may, thermochemical reusing needs acknowledgment as reusing from market and controllers. There are many open inquiries concerning innovation, economy and guideline. With eco-effectiveness examination we guarantee that the inventive methodology makes an incentive for the earth. In addition, substance reusing speaks to an energizing business open door for us and our clients, as the subsequent items are of equivalent quality to the items got from fossil feedstock.

The Advantages of Synthetic Reusing Include

- The procedure results in new synthetic compounds or monomers that can be utilized to make an assortment of new materials, not simply more polyester.
- The waste stream should not be arranged as completely as a waste stream intended for mechanical reusing which possibly works if the waste is completely arranged.
- Contaminants, for example, colours, shades, spandex, and metals can be evacuated or dispensed with amid the concoction reusing process.
- It gives the chance to utilize an alternate impetus amid the polymerization procedure. As of now, antimony trioxide, a speculated cancer-causing agent, is utilized to accelerate the response to make the PET polymer pitch from the monomers.

Mechanically or chemically recycled polyester produced using plastic jugs or clothing contains remaining measures of antimony trioxide. Having dangerous synthetic compounds, for example, antimony trioxide in a material diminishes its feasibility as a round material, since reusing or reuse of that material would propagate the presentation to that lethal compound. The compound reusing process offers the chance to dispose of the antimony trioxide from its unique source (plastic containers or post-buyer waste) and re-catalyse the PET tar with an elective impetus, which is not regular today since antimony trioxide is modest and a powerful impetus [15].

Chemically reused polyester has some natural advantages when contrasted with virgin polyester and since it is, basically, equivalent to virgin polyester, it bodes well that polyester ought to be occupied from the landfill and utilized repeatedly with no misfortune in esteem.

- Synthetically reused polyester can be produced using post-buyer squander, accordingly occupying waste streams from entering the landfill.
- Artificially reused polyester implies utilizes less vitality and water amid PET sap creation.

There are a couple of worldwide organizations with a different scope of developments that address the impetus; a few advancements are in innovative work and others are exorbitant. Industry support is expected to progress artificially reused polyester into circularity and the huge need is to supplant the impetus utilized amid the polymerization process.

Bioplastics from Sustainable Biomass Sources
Reusing is not the best way to diminish plastic's effect on the planet. Most plastics are made of petrochemicals got from non-renewable energy sources, yet another type of bioplastics in the most recent decade depends rather on feedstocks from inexhaustible biomass sources, making them conceivably progressively reasonable. For bio-based plastics there is an expansive range of feedstock choices and numerous headways have been accomplished over the time of a years ago. The first age bioplastics were from starch rich plants, for example, corn or sugar stick. The second era alludes to feedstock not appropriate for sustenance or creature feed. It tends to be either non-nourishment crops (for example cellulose) or waste materials from first era feedstock (for example squander vegetable oil). Additionally, bagasse, corncobs, palmfruit bunces, switch grass. The third era feedstock as of now identifies with biomass from green growth, which has a higher yield than first and second era feedstock. This classification is still in its formative stage. It likewise identifies with bioplastics from CO_2 or methane.

Most bioplastics available today are produced using starches or cellulose, both inexhaustible assets. Sugar stick and corn are the most well-known plants used to create starch or cellulose. Poly acidic corrosive (PLA) is produced using plant material and is biodegradable when in a business fertilizing the soil office. PLA is a biodegradable thermoplastic gotten from inexhaustible assets, most ordinarily cornstarch or sugarcane. Countless plant-based plastics are as of now delivered utilizing a PLA base given that it's promptly accessible and financially savvy. Polylactic corrosive (PLA) is at present a standout amongst the most promising biodegradable polymers (biopolymers). PLA is utilized as a substitution to traditional petrochemical based plastics, essentially as nourishment bundling compartments what's more, movies and all the more as of late, in gadgets and in the production of engineered strands.

Financial Issues Identifying with Reusing
Two key financial drivers impact the suitability of thermoplastics reusing. These are the cost of the reused polymer contrasted and virgin polymer and the expense of reusing contrasted and elective types of adequate transfer. There are extra issues

related with varieties in the amount and nature of supply contrasted and virgin plastics. Absence of data about the accessibility of reused plastics, its quality and appropriateness for explicit applications, can likewise go about as a disincentive to utilize reused material. Verifiably, the essential strategies for waste transfer have been via landfill or cremation. Expenses of landfill fluctuate extensively among locales as indicated by the basic geography and land-use designs and can impact the suitability of reusing as an elective transfer course. In Japan, for instance, the unearthing that is fundamental for landfill is costly a result of the hard idea of the hidden volcanic bedrock; while in the Netherlands it is exorbitant due to porousness from the ocean. High transfer costs are a financial motivation towards either reusing or vitality recuperation.

Accumulation of utilized plastics from family units is increasingly prudent in rural areas where the populace thickness is adequately high to accomplish economies of scale. The most effective accumulation plan can change with territory, kind of homes (houses or huge multi-loft structures) and the kind of arranging offices accessible. Mechanical advances in reusing can improve the financial aspects in two principle ways—by diminishing the expense of reusing (profitability/productivity enhancements) and by shutting the hole between the estimation of reused tar and virgin gum. The last point is especially improved by advancements for transforming recuperated plastic into nourishment grade polymer by expelling sullying—supporting shut circle reusing. Along these lines, while over 10 years prior reusing of plastics without endowments was generally just feasible from post-modern waste, or in areas where the expense of elective types of transfer were high, it is progressively now reasonable on a lot more extensive geographic scale, and for post-shopper squander.

Difficulties and Opportunities for Improving Plastic Recycling
Successful reusing of blended plastics squander is the following significant test for the plastics reusing area. The favourable position is the capacity to reuse a bigger extent of the plastic waste stream by growing post-customer accumulation of plastic bundling to cover a more extensive assortment of materials and pack types. Item structure for reusing can possibly aid such reusing endeavours. Subsequently, more extensive usage of arrangements to advance the utilization of ecological structure standards by industry could large affect reusing execution, expanding the extent of bundling that can financially be gathered and occupied from landfill. A similar rationale applies to tough purchaser merchandise structuring for dismantling, reusing and details for utilization of reused gums are key activities to build reusing.

Most post-customer gathering plans are for unbending bundling as adaptable bundling will in general be risky amid the accumulation and arranging stages. Most present material recuperation offices experience issues dealing with adaptable plastic bundling as a result of the diverse taking care of qualities of inflexible bundling. The low weight-to-volume proportion of movies and plastic packs additionally makes it less monetarily suitable to put resources into the fundamental gathering and arranging offices. In any case, plastic movies are as of now reused from sources including auxiliary bundling, for example, recoil wrap of beds and boxes and some rural movies, so this is possible under the correct conditions. Ways to deal with expanding the reusing of movies and adaptable bundling could incorporate separate accumulation, or interest in additional arranging and preparing offices at recuperation offices for taking care of blended plastic squanders. So as to have effective reusing of blended plastics, superior arranging of the information materials should be performed to guarantee that plastic kinds are isolated to large amounts of immaculateness; there is, be that as it may, a requirement for the further advancement of end markets for every polymer recycle stream.

Future Perspective

The talk of material reusing as a framework cannot be finished up without consideration being paid to the worldwide idea of the framework. Here there is a two-overlay condition: (1) expanded material waste is being made all through the world in light of expanded discretionary cash flow in creating countries. Accordingly, worries for transfer must be considered from all pieces of the world. This has suggestions for multifaceted research. (2) a significant part of the market for utilized garments is situated in creating nations where yearly wages are once in a while not exactly the expense of one outfit at retail cost in the United States. The creating nation markets give a setting where exceedingly industrialized countries can change their over the top utilization into a valuable fare. For huge numbers of these individuals, utilized dress surplus gives a truly necessary administration. Shockingly, worldwide exchange laws frequently hamper the free progression of utilized garments. By raising cognizance concerning natural issues, channels for transfer, and earth cognizant business morals, steps can be made toward an increasingly practical condition. Resident concerns campaigned with districts will likewise expand the quantity of regions that offer material reusing as one of the classes of their waste administration process. To reuse effectively, shoppers must grasp the framework, not simply make an intermittent beneficent gift. In the interim judges must keep on growing new esteem markets and market the after-use conceivable outcomes so the framework capacities at full limit and with responsibility from all.

References

1. Kalliala E, Nousiainen P (1999) Life cycle assessment: environmental profile of cotton and polyester/cotton fabrics. AUTEX Res J 1(1):8–20
2. Mueller R-F (2006) Biological degradation of synthetic polyesters—enzymes as potential catalysts for polyester recycling. Process Biochem 41:2124–2128
3. Utebay B, Celik P, Cay A (2019) Effects of cotton textile waste properties on recycled fibre quality. J Clean Prod 222:29–35
4. Filho WL, Ellams D, Han S, Tyler D, Boiten VJ, Paco A, Moora H, Balogun A-L (2019) A review of the socio-economic advantages of textile recycling. J Clean Prod 218:10–20
5. Zambrano MC, Pawlak JJ, Daystar J, Ankeny M, Cheng JJ, Venditti RA (2019) Microfibers generated from the laundering of cotton, rayon and polyester based fabrics and their aquatic biodegradation. Mar Pollut Bull 142:394–407

6. Raheem AB, Noor ZZ, Hassan A, Hamid MKA, Samsudin SA, Sabeen AH (2019) Current developments in chemical recycling of post-consumer polyethylene terephthalate wastes for new materials production: a review. J Clean Prod 225:1052–1064
7. Park SH, Kim SH (2014) Poly (ethylene terephthalate) recycling for high value added textiles. Fash Text 1(1):1
8. Blackburn R, Payne J (2004) Life cycle analysis of cotton towels: impact of domestic laundering and recommendations for extending periods between washing. Green Chem 6(7):59–61
9. Hopewell J, Dvorak R, Kosior E (2009) Plastics recycling: challenges and opportunities. Philos Trans R Soc London B Biol Sci 364(1526):2115–2126
10. Al-Salem S, Lettieri P, Baeyens J (2009) Recycling and recovery routes of plastic solid waste (PSW): a review. Waste Manag 29(10):2625–2643
11. Thakur S, Verma A, Sharma B, Chaudhary J, Tamulevicius S, Thakur VK (2018) Recent developments in recycling of polystyrene based plastics. Curr Opin Green Sustain Chem 13:32–38
12. Ragaert K, Delva L, Geem KV (2017) Mechanical and chemical recycling of solid plastic waste. Waste Manag 69:24–58
13. Delva L, Hubo S, Cardon L, Ragaert K (2018) On the role of flame retardants in mechanical recycling of solid plastic waste. Waste Manag 82:198–206
14. George N, Kurian T (2014) Recent developments in the chemical recycling of postconsumer poly(ethylene terephthalate) waste. Ind Eng Chem Res 53(37):14185–14198
15. Liu W, Liu S, Liu T, Liu T, Zhang J, Liu H (2019) Eco-friendly post-consumer cotton waste recycling for regenerated cellulose fibers. Carbohyd Polym 206:141–148

Recycled Polyester—Tool for Savings in the Use of Virgin Raw Material

Shanthi Radhakrishnan, Preethi Vetrivel, Aishwarya Vinodkumar and Hareni Palanisamy

Abstract The demand for Polyester is rising especially in the Asia Pacific region as predicted by the textile forecast. The superior quality, established manufacturing and lower economics has made it a practical alternative for other fibers. The possibility of recycling has created a growing awareness for the use of RPET for the manufacture of sustainable clothing. Further polyester fibers can be made functional by incorporating special properties of hi-tech nature opening new opportunities in the global market. Sportswear and home textiles are two major segments that capitalize in the use of polyester fibers. The sustainability drive looks for the environmental footprints by using tools like life cycle assessment of products and materials. The human demand on nature for the growth and development of a region/product or process is inevitable; the capacity of nature to regenerate will be spoken in terms of environmental footprint. Virgin polyester has a greater environmental footprint when compared to RPET as reiterated by various case studies and reports. This paper will highlight the environmental foot print of RPET and virgin polyester with suitable examples to understand the basis for the use of recycled polyester. A case study has been analyzed to show the benefits of using Recycled hollow polyester for functional pillows. Leading retailers in the apparel sector communicate to the consumers the percentage of RPET in their apparel through labels and advertisements to promote sustainability awareness ending in product sales. The future is recycling and reuse and RPET will play an important role in reaching the targets and aspirations of the textile world.

Keywords Closed loop recycling · Open loop recycling · Virgin hollow polyester · Recycled hollow polyester · Functional pillows

S. Radhakrishnan (✉) · P. Vetrivel · A. Vinodkumar · H. Palanisamy
Department of Fashion Technology, Kumaraguru College of Technology, Coimbatore, India
e-mail: shanradkri@gmail.com

© Springer Nature Singapore Pte Ltd. 2020
S. S. Muthu (ed.), *Environmental Footprints of Recycled Polyester*, Textile Science and Clothing Technology, https://doi.org/10.1007/978-981-13-9578-9_3

1 Introduction

Polyester is an all time favorite of many consumers due to its inherent properties like durability, wrinkle resistance, quick drying nature and easy maintenance. It is a popular choice in fashion as it provides all the qualities essential for apparel and is easily available for producing all types of fabrics from casual wear to formal/functional wear due to its ability to blend with any type of fibers. This fiber was invented by the English researchers of the Imperial Chemical Industries in 1940 and DuPont purchased the rights and started production of Dacron Polyester in 1953 [1]. The fiber was considered as a wonder fiber and was advertised as a 'miracle fiber that could be worn for 68 days straight without ironing and still look presentable' [2]. The fiber has gone through various stages of transformation to reach the high tech market that brought about a major revolution in active sportswear and swim wear. The endorsements from the Manmade Fiber Association and the Council of Fashion Designers with members like Oscar de la Renta, Mary Mcfadden, Perry Ellis and Calvin Klein, helped to boost the image of polyester among the public.

In 1958, a new polyester fiber Kodel was developed by Eastman Chemical Products, Inc. The polyester market grew by leaps and bounds and the industry expanded rapidly till 1970. The badly styled double knit suit [3] brought the downfall of the polyester image and was termed as the uncomfortable fabric and shunned by all. Viscose and natural fibers were opted for their comfort and coolness. The change in the consumers attitude to look out for something natural brought about the demand for cotton. Research and development have brought about new forms of polyester in the 1990s. In 1991, the emergence of polyester luxury fibers changed the trend and the industry is experiencing revival. Micro fibers enhanced the feel of polyester to resonate silk [4]. US designer Mary Mcfadden created her line of fashion garments with the new form of polyester. The North Carolina State University has undertaken research to develop a strong polyester fiber equivalent to Kevlar for bullet proof vests [4]. Studies conducted by Hoechst Fibers between 1981–83 [5] showed that 89% of the consumers were not able to differentiate between polyester and natural fibers; they were interested in the appearance of the apparel and not in the content. Polyester made from microfibers had the feel of silk which accounted for the expensive tag. This marked the new era of the polyester image.

1.1 Forecast for Polyester

The growth rate of polyester fiber consumption is around 7% per year since the year 1990 and it occupies about half the total consumption of global manmade and natural fibers. In 2017, Polyester yarn accounted to 69% of the total yarn consumption; apparel manufacture and home furnishing sector were the major users of polyester as raw material. The global fiber production increased to 53.7 million metric tons [6] and about 94% of polyester fiber production on a global scale has shifted to Asia.

The forecast for 2022 states that Northeast Asia will be center for polyester fiber covering about 80% of facility and 75% of the demand growth. India will be the second largest producer followed by Southeast Asia. The polyester fiber production and consumption will be limited in the remaining parts of the world while Asia will remain the hub for polyester production. The polyester fiber market is expected to grow on an average of 1–3% annually [7]. Reports also state that the growth of polyester fibers will be 4% in tune with the growth rate of the GDP of the emerging world.

1.2   Impact of Polyester Fiber Production

Petrochemicals form the basic raw materials for polyester leading to very strong non-biodegradable fabrics. Cooling is an important process in polyester manufacture which uses large amounts of water with lubricants. It has been reported that 70 million barrels of oil [8] is used every year for polyester production. A life cycle assessment report presented by two major Scandinavian textile service companies and a Finnish textile manufacturing company (wert processing of cotton and polyester/cotton) focusing in hotel textiles, state that polyester fiber production use 40% more energy than cotton fiber production while the consumption of water is far more less (17 L per kg) in the case of polyester fiber production when compared to (7–29 m^3/kg—conversion factor [1 m^3 = 1000 L] [9] cotton production [10–12]. These processes may be very polluting in nature and energy driven. Polyester manufacture requires double the energy required for the manufacture of conventional cotton. Carcinogens are harmful chemicals that create problems to the health of living organisms. Polyester is manufactured in developing countries like China, Indonesia and Bangladesh where environmental regulations are not serious and sloppy leading to untreated discharge causing impairment in health to people who inhabit the downstream areas and near the manufacturing industries. The consumption of water for the production of polyester is lower than the amount used for natural fiber manufacture. It is simpler to dye Polyester in the solution stage before the fibers are produced; dyeing of polyester in the latter stages requires high pressure and the presence of a carrier. A few disperse dyes used for dyeing polyester are allergic in nature and result in contact dermatitis [13]. Synthetic dyes are used to dye polyester and the dye remains that are left untreated pollutes river bodies killing aquatic and wildlife and also poisoning individuals who depend on water bodies in many developing countries.

Non renewable resources like coal and petroleum [14, 15] are used as raw materials for the manufacture of polyester which are extracted through mining. Mining covers a large area and the natural habitat is destroyed in the process. The conversion of raw materials to finished products involves transportation which adds to the environmental foot print. The energy requirement for the production of polyester is twice that used for conventional cotton is increases considerably when compared to other fibers like organic cotton, hemp etc. Further the amount of polyester produced is high and this will culminate in more transportation impacts. The byproducts of polyester are negligible and hence sustainable. Packaging is excessive for safety of goods, which ends in landfills. The disposal of polyester goods depends on the type of product e.g. when combined with nylon the decomposition time may be 30–40 years.

The microfiber pollution of the oceans is an invisible problem that has caused heavy pollution of water bodies that receive the laundry wash liquid. The microfibers released from textiles and apparel into the marine environment is estimated to be 0.19 million tons annually [15, 16] and every time a polyester or synthetic apparel is washed, around 2000 micro fibers (1.7 g) are released into the drain [8, 17]. A team of researchers from the University of Barcelona have established that the oceans of Southern Europe from the Cantabrian Sea to the Black Sea house large quantities of microfibers which are within the range of 3–8 mm length and less than 0.1 mm diameter extruded from home and industrial washing machines [18–20]. A report also states that 50 mL of sediment from the sea floor of Southern European seas, have 10–70 microfibers [21] while 20% of the microfibers are found at the depth of 2000 m in the open seas which is ingested by the aquatic living organisms. A study reported in 2017, states that the knit structure of textiles play a vital role in shedding of micro fibers during washing; It has been highlighted that the greatest amount of microfiber shedding during washing was polyester fleece fabric (7360 fibers/m^{-2}/L^{-2}) when compared to the other polyester knits (87 fibers/m^{-2}/L^{-2}) (RP-43). The release of microfibers is 3.5 times more in the case of tumble drying when compared to the washing process [22]. An analysis conducted with beach sand scooped from 18 different beaches across six continents followed by filtration and chemical analysis showed that 80% of the fibers were made of polyester [23]. Microfibers released from synthetic textiles and apparels are readily consumed by the fish and other wildlife as they are very fine; they tend to bioaccumalate producing toxins in the bodies of smaller and larger animals in the food chain [17]. This problem can be solved by using biosynthetic fibers, high twist yarns, avoiding slack textile construction and incorporating a filtering mechanism in the washing machine.

1.3 Recycling of Polyester

The conversion of waste materials into new raw materials for production or into new products for consumer use is recycling. This system promotes reduction in the use of virgin materials for manufacture thereby saving energy and avoiding air and water pollution. In the waste hierarchy, 'recycle' is the third important component after 'reduce' and 'reuse' [24, 25]. Recycled materials and products are available in the market and to standardize the process/product, ISO standards for plastic waste (ISO 15270:2008) [26] and environmental management control of recycling (ISO 14001: 2015) [27] have been established. The concept of recycling will lead to the formation of many more such standards as a result of research and product innovation.

In the US about 4 million plastic bottles are disposed every hour [28] leading to environmental problems. This problem has been addressed by recycling the empty spent bottles into recycled polyester fibers.

1.4 Characteristics of Recycled Polyester

Textile Exchange, a non profit organization has confronted about 50 retailers and apparel companies to incorporate at least 25% of recycled polyester in their product line by 2020 but by 2019 the use of recycled polyester has increased up to 36%. The forecast by Textile Exchange for 2030 is that 20% of the total polyester produced will be recycled. Many studies reveal that the properties of virgin and recycled polyester are similar except for strength which reduces in the case of recycled polyester due to the recycling process. This may be so when the polyester is recycled for the first time but as it passes through a number of recycling process the properties will vary to a greater extent.

While analyzing the properties of virgin and recycled polyester in Table 1, the crystallinity of recycled polyester is 13.85% lower than that of virgin polyester which may be accounted to the process of remelting and reorganization of the polymer in the recycling process. The difference in the glass transition property between virgin and recycled polyester fiber is attributed to the chemical input during the recycling process, leading to uneven dyeing behavior of recycled polyester. The differences in the thermal properties is due to the cross linking and reformation of the molecular chains due to the remelting of polyester. Presence of other materials due to recycling process, is highlighted due to a higher residue in the case of recycled polyester fiber.

The strength properties is based on the molecular weight of the filaments which is founded on the intrinsic velocity approach. Due to the recycling process polymer degradation occurs leading to the presence of smaller polymeric chains. Longer the chains stiffer is the material characteristic and higher the intrinsic viscosity [31]. Fibers with lower molecular weight will show lower breaking stress and break elongation and higher Young's modulus [32–36]. The extrusion process in the manufacture of recycled polyester culminates in hydrolytic degradation caused by the difference in flow index of the raw materials used in the process. Increase in melt flow shows the thermal-mechanical degradation of the material [36]. Lower shrinkage percentage is due to the alignment of molecules after the recycling process. With regards to the fiber orientation, the polarized attenuated total reflectance infrared technique showed that virgin fibers have better performance than the recycled ones [31]. The pros and cons of using recycled polyester is given in Table 2 [37].

Table 1 Comparitive analysis—properties of virgin and recycled polyester [29, 30]

S. No.	Parameters	Virgin polyester	Recycled polyester
1.	*Crystallinity*		
	Crystalline region	37.2	28.5
	Amorphous region	62.8	71.5
2.	*Thermal properties*		
	Glass transition property (°C)	75.35	87.27
	Melting point	242.72	244.15
	Degradation temperature	419.99	419.62
	residue	0.79	10.80
3.	Average molecular weight (g mol)	19,342	15,812
4.	*Strength properties*		
	Tensile strength (kg/cm^2)	140.5	220.0
	Breaking stress (kg/cm^2)	82.2	42.2
	Elongation at break (%)	6.96	5.00
	Young's modulus (kg/cm^2)	5690	10,500
5.	Shrinkage % (150 denier)	9.31	6.18

Table 2 Pros and cons of using recycled polyester

Advantages	Disadvantages
• Prevents plastics from polluting the environment • rPET has properties almost similar to virgin polyester and is reusable in the manufacturing stream • Life cycle assessments prove that there is a reduction of carbon emissions and prevention of toxic emissions due to incineration • Promotes research for developing new technologies to develop a non virgin supply chain to include rPET in the production cycle	• Polyester blended with other fibers are difficult to recycle • 100% polyester cannot be recycled infinitely. Each time the material is recycled the polymer gets degraded and can be used for low quality materials or mixed with some quantity of virgin polyester • PET bottles leach antimony which causes cancer • Contribute to microfiber pollution in oceans and water bodies • rPET is formed from a mixture of waste bottles and hence the base colour of the raw material. To rectify this discrepancy chlorine based bleaching is undertaken • Raw material color variation leads to inconsistency in dye shade. Redyeing is followed which results in high water, energy and chemical use

1.5 Recycling Techniques

An estimate states that PET Bottle recycling was 13 million tons in 2018 and is expected to rise up to 15 million tons by 2020 [38]. Environmental concerns have brought PET bottle waste recycling to rPET fibers into the commercial limelight. The recycled PET flakes obtained from PET bottles are utilized for the production of staple polyester fiber and used as raw material in the textile sector. Recycling of materials can be performed by two methods namely the closed loop recycling and the open loop recycling. In the closed loop system the post consumer waste is recycled to be used in the same production system e.g. recycling of plastic cans for production of plastic cans while the open loop system recycles the post consumer waste to be used as raw material for another production system. Eg. recycling of plastic cans for production of recycled polyester fiber [39–43]. Chemical recycling and mechanical recycling are commonly used to recycle PET bottle to polyester fibers as in Fig. 1.

Chemical recycling produces superior quality material but is labor and power intensive leading to higher costs when compared to mechanical recycling [45]. Consistency in the quality of the rPET fiber to equal virgin fiber can be achieved by taking special care during the sorting and the cleaning processes. There are many case studies where different recycling techniques have been used e.g. CARBIOS, France uses enzymes to Biorecycle PET; Gr3n uses microwave reactor for depolymerization stage in PET recycling; Loop Industries recycling process uses selective solvents to procure high quality PET; Designtex has developed Loop-to-Loop recycling by producing good quality PET yarns from recycled feedstock [46–49]. This list can go on but it shows the intensity of the environmental safety by reuse of waste PET by means of chemical or mechanical recycling.

Researchers from the Hongkong Research Institute of Textiles and Apparels (HKRITA) have developed a patented chemical solution where the cotton in the blended textile or apparel is broken down into cellulose in the presence of heat and water and can be separated from the fabric; the remaining polyester fibers are converted into garments without reduction in quality. This concept is termed as fibre-to-

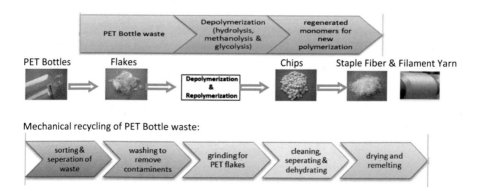

Fig. 1 Chemical and mechanical recycling of polyester [44]

fibre recycling. Closed Loop Apparel Recycling Program, jointly partnered by H&M Foundation and HKRITA aims to recycle apparels made from blended textiles to be commercialized by 2020 [50].

Patagonia, US based company manufacturing active sportswear, have incorporated the use of rPET in their sportswear. They encourage return of garments after the full use of the clothing and advise customers to buy only when in need. Outdoor jackets and 'Riz Boardshorts' for surfing are made of 100% recycled polyester which can be returned with discounts and for reuse.

Kate Goldsworthy, a zero waste UK designer has developed Zero waste polyester dress, mono finished surface finishes using laser instead of traditional notions for value addition of garments. As glues, threads, stitches, chemicals and pastes, are not used the garment can be recycled without any quality loss [44]. The concept of 'Twice Upcycled' has been developed by Earley and Goldsworthy. In the first stage a second hand polyester shirt is redefined with overprints and reshaping for use by consumer. The shirt is returned after wear for the second upcycling where the shirt is recut used as lining for a recycled polyester fleece laser welded quilted jacket [51]. Similar projects have been undertaken as 'Top 100', Teijin Eco Circle, Patagonia Common Threads and Closed-Loop Recycling.

1.6 Environmental Footprint of Recycled Polyester

Environmental footprint takes into account the impact of an individual/population/activity on 'biologically productive land' and 'sea area'. All manufacturing and production practices is required to produce all the resources consumed for the activity and take responsibility to digest the waste generated due to the activity, with the help of technology and resource management schemes. Before going on about measuring Environmental or Ecological Footprint we need to understand the term 'biocapacity'. Biocapacity is another term for biological capacity, the ability of an eco system to produce biological resources and absorb carbon dioxide emissions. 'Ecological overshoot' is a condition when the population's demand exceeds the capacity of the ecosystem to regenerate the resources consumed by humans and to absorb carbon dioxide emissions [52].

To estimate the Ecological Footprint of a region or nation, the ecological footprint of the individual is calculated as mentioned below:

- [A] = The amount of material consumed per person as tons per year − the person's demand for biologically productive space (e.g. cropland/water requirement to grow materials or to sequester CO^2 emissions)
- [B] = The yield of the specific land or sea area as annual tons per hectare
- [A/B] will give the number of hectares per person

This result is converted to global hectares using yield and equivalence factors. A global hectare is the biological productivity of a hectare with world average productivity [53].

- Total Ecological footprint per person is sum of global hectares (of all activities) required to support the person
- This value is multiplied by the number of persons to get the ecological footprint of a nation/country etc.

The average global ecological footprint in 2012 was 2.84 global hectares per person, while the world average bio capacity is 1.73 global hectares per person; the global ecological deficit is 1.1 global hectares per person [52, 53].

This means that humanity uses the equivalent of 1.7 Earths for provision of resources and absorption of the waste generated; it also states that the regeneration period for the Earth is one year six months for the ecological resources and services we take from nature annually. Human activity by over production due to demand [e.g. fishing and harvesting from forests] is emitting more CO^2 into the atmosphere than the sequestering capacity of the forests.

To reduce the ecological footprint, lot of efforts has been taken to use technology and research to modify and streamline processes in the industry and in society. The Swedish Environmental Protection Agency states that between the years 2000 and 2009 the textile waste increased by 40%. Mono fiber apparel is easier to recycle than apparels made from blends. Waste consisting of 50% cotton and 50% polyester were taken as samples and three different recycling techniques were employed; the environmental performance was assessed by life cycle assessment [LCA]. The processes compared were material reuse, separation of cellulose from the blended fabric using solvent (N-methylmorpholine-N-oxide) and incineration which was the conventional process followed in Sweden. The best results were obtained from the material reuse process where the savings estimated was 8 tons of CO_2 equivalents and 164 GJ of energy per ton of textile waste. An integration of the recycling techniques according to the suitability of materials would save around 10 tons of CO_2 equivalents and 169 GJ of energy per ton of textile waste [54].

In the UK, one million tons of clothes are disposed into the domestic waste flow annually. Collection services exist for reuse and reprocessing of textile waste. The Salvation Army Trading Company Limited (SATCOL), UK, has collection and distribution center for donated clothing and shoes. A study had been conducted to estimate the energy requirement for retailing and distribution of the donations by reuse in comparison with the manufacture from virgin materials. The results showed that in the case of reuse, one ton of polyester garment used 1.8% of the energy required for the manufacture of the same from virgin materials while one ton of cotton clothing used 2.6% of the energy required for the manufacture from virgin materials [55]. Similarly many studies reveal savings and betterment to the environment when reuse and recycling is undertaken and recycled materials enter the production stream as raw materials for the next production cycle. To understand the importance of recycling and its benefit a case study has been discussed.

2 Case Study: Design and Development of Antimicrobial Functional Pillows Using Recycled Hollow Polyester

2.1 Introduction

Pillows are supporting tools for the body and helps in alleviating common forms of joint pains. During sleep, pillows maintain the upper body alignment, relieve pressure and counter balance the different points in the body. Therapeutic pillows are used after surgery and act as reinforcements and supportive strengthening materials. After joint replacement surgeries, pillows are judiciously placed for desired alignment and healing and pain relieving aids. Pillows are available today for all types of functions and are termed as 'functional pillows' which serve both the resting function and therapeutic function.

Pillows are filled with all types of fillers—natural and synthetic in nature. This project deals with the synthetic filling materials used in pillows which have resilience, springiness, strength and flexibility. Synthetic filling is known for durability, comfort, machine wash ability and ease of care and maintenance. The most common synthetic filling material found in India is hollow polyester fibres. Batting material is considered as a hollow fiber when the fiber has a space in between. The difference between a polyester fiber cross section and a hollow polyester fiber cross section is given in Fig. 2.

Thermal comfort is dependent on water vapor permeability, thermal insulation and the water transport of the material. In the case of hollow fibers the central porous region is responsible for better transport and vapor permeability leading to better comfort [57]. The hollow fibers used for pillow filling is mechanically crimped to incorporate extensibility and to accommodate compression [58]. The researchers in the University of Manchester, Woodcock's research team, conducted a study by analyzing filling samples from ten pillows used for 18 months to 20 years. This study revealed that there were few thousand spores of fungus per gram of sample and the

Parameters	Polyester Fiber	Hollow Polyester Fiber
Spinneret hole design	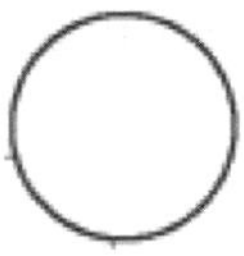	
Fiber Cross section		

Fig. 2 Cross sectional view of polyester fibers [55, 56]

fungal species ranged from four to sixteen in number. The most common fungus found especially in synthetic fiberfill pillows was Aspergillus fumigates [59]. The pillow becomes a miniature eco system for the growth of fungal species. Pillows are meant to be changed after 2–3 years of use but not many consumers follow this directive. If proper care is not followed many would be burying their heads into a bed of fungal spores while sleeping.

Raw material plays an important role in exploiting the environmental footprint. Virgin materials are being replaced by recycled materials not only for the cost effectiveness but also because of the environmental impact. According to the report issued by The Guardian one million plastic bottles are used every minute and is expected to rise by 20% by 2021 forming a reasonable basis to global warming [60]. A Hindustan Times report states that India recycles 90% of the PET waste generated when compared to the lower recycling percentage of PET waste in other countries like Japan (72.1%), Europe (48.3%) and US (31%) [61]. The remaining percentage ends up in landfills. The waste PET bottles can be used to produce Recycled Polyester Fibre termed as RPET. The Ministry of New and Renewable Energy, Government of India offers financial incentives for plastic recycling plants as interest rate reduction to 7.5% capitalized with an annual discount rate of 12% [62]. The most important emerging issue in the present phase of industrialization is manufacturing environment friendly products in a sustainable way—circular economy. The concept of recycling is considered the most sustainable practice in the industrial sector. This study focuses on the selection of environment friendly sustainable raw material [rPET] from PET bottles, for the development of functional pillows with the special utility of anti bacterial finish. This background served as a base for the research study with the following objectives

- Designing of functional pillows
- Functional Pillow Development using Recycled hollow polyester and Antimicrobial functional finish
- Analysis of impact of using recycled materials in product development.

2.2 Materials and Methods

Materials: The materials chosen for the study is given in Table 3.

2.2.1 Fiber Selection for Pillow Filling

Three fibers were selected for the study. Kapok is a commonly used filling material naturally available in India. For the sake of comparison, properties of Virgin hollow Polyester, Recycled Hollow Polyester and Kapok was studied.

Table 3 Materials chosen for the case study

Materials	Option 1	Option 2	Option 3
Fiber selection for pillow filling	Virgin hollow polyester	Recycled hollow polyester	Kapok
Yarns used for pillow shell development	100% recycled cotton	90:10 recycled cotton: recycled polyester	80:20 recycled cotton: recycled polyester
Fabric development blend ratios used	100% recycled cotton	90:10 recycled cotton: recycled polyester	80:20 recycled cotton: recycled polyester

Virgin Polyester

Virgin polyester was sourced from Reliance Pvt Ltd, Coimbatore. A market study was done and it was found that most commercially available pillow stuffing was Recron. These fibers have the properties of softness and resilience for a long time.

Recycled Hollow Polyester from PET Bottles

Recycled hollow polyester fiber obtained from used PET bottles was sourced from Sulochana Fibres, Tirupur. The process of recycling reduces the strength and color of the polyester but all other properties are similar to virgin polyester. These fibers were selected to focus on sustainability and reduce the use of virgin material which has a negative impact on the environment in terms of resource utilization, pollution and green house gases.

Kapok

Kapok fiber is one of the natural cellulosic fibers which grow on the kapok plant. It has a hollow body and a sealed tail, which is traditionally used fibre for pillow stuffing. It gives more comfort and durability.

2.2.2 Testing of the Fibers

The fibers were tested for Length and Denier, Tenacity, Crimp, Thermal Conductivity and Water Vapor Permeability

2.2.3   Selection of Filling Fiber Blend Ratio

Three different blend ratios namely 100% Virgin Hollow Polyester, 100% Recycled Hollow Polyester and Virgin Hollow Polyester:Recycled Hollow Polyester 50:50, were selected for the study.

2.2.4   Selection of Yarn and Blend Ratio for Pillow Slip Fabric

Recycled polyester and recycled cotton blended yarn was sourced from Ganapathy Mills Tirupur. Fabric made from poly cotton blend combines the strength of two fibers. The recycling process reduce the strength of the yarn but it maintains the sustainability aspect by avoiding the manufacturing of virgin fibers. Three different types of blend ratios were selected for yarn manufacturing namely 100% Recycled Cotton, 90:10 Blend of Recycled Cotton and Recycled hollow polyester and 80:20 Blend of Recycled Cotton and Recycled hollow polyester.

2.2.5   Testing of the Yarns for Pillow Slip Fabric

The yarns were tested for Count and Lea Strength, Single Yarn Twist (ASTM D 1423), Yarn Appearance Grade and Blend Ratio estimate.

2.2.6   Fabric Development for Pillow Slip

Plain weave fabric was woven using Shakti 2500 Power Loom Machine in Ganapathy Mills, Tirupur as shown in Figs. 3 and 4. The machine particulars are given in Table 4. The setting of the fabric was EPI 52; PPI 40 and the width 52″.

Fig. 3 Weaving of fabric

Fig. 4 Shuttle with pirn and cones

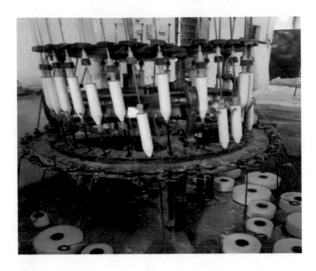

Table 4 Shakti 2500 power loom machine parameters for fabric development

Frequency	50 Hz
Voltage	220–240 VAC
Power consumption	Up to 5 kw
Material	Mild steel
Speed	Up to 200 RPM
Max shedding	50–120 mm

2.2.7 Testing of Pillow Slip Fabric

The fabric produced was tested for GSM and Thickness, Tear strength, Thermal Conductivity, Air permeability, Bending length, Water vapor permeability, Crease Resistance, Bursting strength and Drape.

2.2.8 Optimization Using Box-Behnken Design of Experiment

In the Box and Behnken factorial design the variables are selected at three levels, viz. +1, 0 and −1. The response Y is given by a second-order polynomial, i.e.

$$y = \beta_0 + \sum_{j=1}^{k} \beta_j X_j + \sum_{j=1}^{k} \beta_{jj} X_j^2 + \sum_{i=1}^{j-1} \sum_{j=2}^{k} \beta_{ij} X_i X_j + \varepsilon$$

where, X, is the i variable, K is the number of variables and b_0, b_1, and b_2 are the regression coefficients associated with the variables. To find out the regression coefficient, the response Y has to be found out by using different experimental combinations of the variables under consideration.

Table 5 Levels of variables and factors chosen for the study

Variables	Factors	−1	0	+1
A	Fiber ratio	100% RHP	50:50 RHP:VP	100% VP
B	Weight of the fiber (g)	450	500	550
C	Fabric	80:20 RC:RP	90:10 RC:RP	100% RC

Fig. 5 Samples developed for the Box and Behnken statistical design

The factors taken for Box-Behnken Design of Experiment were the filling fiber ratio, weight of filling fiber and the yarn blend ratio were taken at three different levels (+1, −1 and 0) as mentioned in Table 5 fifteen different combinations were taken to optimize the parameters for functional pillow development. The pillow sample is developed for 13 different combination as shown in Fig. 5. The dimension of sample pillow was taken as 1/4th measurement of the standard pillow dimension which is 20″ * 26″ [63]. Three responses 'Y' were taken as thermal conductivity, air permeability and water vapour permeability to optimize the parameters for the pillows.

2.2.9 Anti-microbial Finish for Fabric and Filling

Clothing and textile materials are not only the carriers of microorganisms such as pathogenic bacteria, odour generating bacteria and mould fungi, but also good media for the growth of the micro organisms. Microbial infestation poses danger to both living and non-living matters. Anti-microbial finish was given to prevent the growth

Fig. 6 Chemicals used for
processing

Fig. 7 Process sequence of
anti bacterial treatment

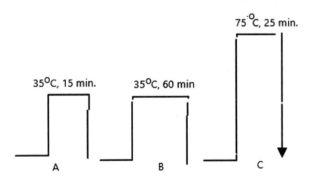

of dust mites and ensure the safety of the person. The anti microbial agent chosen
was N9 Pure Silver™—Micro PD sourced from N9 World Technologies Pvt. Ltd.,
Karnataka, India, Fig. 6. The recipe for the filling fiber and the pillow slip fabric is
given in Table 6. The flow process for the fiber and fabric is given in Fig. 7.

The fiber/ fabric was loaded into the Ramsons drum washing machine with the
required quantity of water and wetting agent and rotated for 15 min at 35 °C [A].

Table 6 Recipe for processing filling fiber and pillow slip fabric

Fabric		Filling fiber	
Water	50 L	Water	50 L
Wetting agent (Spiro wet N)	3 gpl	Ultrafab UPE (Wicking)	10 gpl
Time	15 min	Time	15 min
Temperature	35 °C	Temperature	35 °C
N9 Pure Silver™—Micro PD	10 gpl	N9 Pure SilverTM—Micro PD	10 gpl
pH	5.5/6	pH	5.5/6
Material wt	15 kg	Material wt	15 kg
MLR	1:3	MLR	1:3

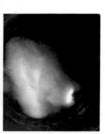

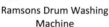

Ramsons Drum Washing Machine | Hydro extractor | Ramsons dryer | Antibacterial treated filling fiber

Fig. 8 Equipment used for processing

This was followed by the addition of antibacterial agent and rotated for 60 min [B], hydro extraction and dried in a Ramsons tumble dryer for 25 min [C]. Figure 8 show the different stages of the process.

2.2.10 Testing the Efficacy of Antimicrobial Finish

JIS L 1902 is an antimicrobial product test for testing antibacterial activity and efficacy on textiles. It was originally developed as a Japanese standard but now used for all Asian countries. The JIS L 1902 method has since been adopted more or less intact as an International Standards Organization (ISO) method, ISO 20743. This method is frequently used for basic fabrics and textiles used in bedding, furniture and apparel. The quantitative approach requires additional incubation time period for bacteria determination by plating. Selected were microbes most popular in testing: Gram (−) *Escherichia coli* ATCC 11229 and Gram (+) *Staphylococcus aureus* ATCC 6538. AS PER JISL 1902, the criteria for bactericidal activity should be not less than 0 and bacteriostatic activity should be not less than 2.

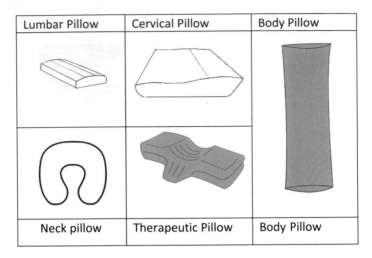

Fig. 9 Selected functional pillows

2.2.11 Design and Development of Functional Pillow

Five functional pillows were designed to suit the different parts of the body namely Lumbar pillow, neck pillow, cervical pillow, therapeutic pillow and body pillow. The designs are given in Fig. 9. The outer pillow slip was made from recycled cotton, the 100% recycled hollow polyester was used for the filling of the pillow and the next choice for therapeutic pillow was thermal bonded batted polyester made from recycled hollow polyester and low melt polyester in the ratio 90:10. The measurements were defined, the patterns for each component of the pillow were drafted and the construction procedure was determined to enable the production of the functional pillows.

2.2.12 Statistical Analysis

The statistical study which measures the differences among group means is analysis of variance (ANOVA). In the simplest form ANOVA is a statistical test to analyze whether or not the means of several groups are equal [64]. ANOVA testing is useful for comparing three or more means (groups/variables), for statistical significance. In this study the samples A, B and C have been compared with the control group as well as between the groups. ANOVA is represented by the F value, when greater than five there is significant difference between the means analyzed. Statistical significance is found to be prominent when the p-value is less than 0.05 which states that all means are not equal or differences are present between the means compared. All the properties studied were tested for ANOVA and the confidence level was set at 95%.

2.2.13   Saving Analysis

Cost/benefit analysis is estimation and evaluation of net benefits associated with alternatives for achieving defined goals of the business and is the primary method used to justify expenditures. Usually the analyst sums the benefits of a project or action and sub tracts the costs to study the feasibility of the project. The study aims at highlighting the benefits of using recycled material in product development. The cause summary is given as its constituent parts for virgin material and compared with recycled material. The individual cost of all five functional pillows is calculated and compared with virgin and recycled raw material. The cost saving analysis is determined in terms of price and production of small scale industries having a turnover of 20 lakhs per annum.

2.3   Results and Discussion

2.3.1   Test Results of the Selected Fibers (Virgin Hollow Polyester, Recycled Hollow Polyester and Kapok)

The physical properties of the selected fibers are given in Table 7. A comparison was made with Kapok as it is commercially used pillow filling in India. The comfort properties are given in Fig. 10.

From Table 7, it can be understood that the denier of recycled hollow polyester is 7.51% lesser than virgin polyester while the denier of Kapok is 29% lesser than virgin polyester. Higher denier polyester (1200) is denser and thicker than the lower denier fiber (600). The crimp of recycled hollow polyester is 33% higher than virgin polyester while the crimp of Kapok is 67% lesser than virgin polyester; highly crimped fibers have a higher resilience. The tenacity of recycled hollow polyester and kapok is lower than virgin polyester by 36 and 58.5% respectively indicating

Table 7   Physical properties of virgin hollow polyester, recycled hollow polyester and kapok)

Parameters	Standard VHP (100%)	RHP (100%)	% gain or loss over original	Kapok	% gain or loss over original	F value
Denier	15.43	14.27	−7.51	10.94	−29.09	677.54*
Crimp (%)	3	4	33.33	1	−66.66	772.81*
Tenacity (g/den)	3.38	2.47	−36.84	1.4	−58.57	594.68*
Length (cm)	9	6.4	−28.88	1	−88.88	3876.2*
Elongation	43.36	35.8	−17.43	4.23	−90.24	64.29*

VHP Virgin hollow polyester; RHP Recycled hollow polyester
*Significant at 5% level

that in the case of recycled hollow polyester, the recycling process has reduced the strength of the material. The length of recycled hollow polyester is 28.88% lesser than virgin polyester and the length of Kapok is 88.88% lesser than virgin polyester; the recycling process may have reduced the length of fiber. Higher the length of the fiber, lesser the amount of fiber required to fill the pillow. The elongation of virgin polyester is 17.43% higher than the elongation of recycled hollow polyester and 90.24% higher than Kapok; the recycling process has reduced the stretch ability of the material in case of recycled hollow polyester. Hence it can be understood that virgin polyester proves to be better than recycled hollow polyester and kapok in terms of denier, tenacity, length, and elongation. The crimp of the recycled hollow polyester is better than virgin polyester and kapok. The F value for all parameters shown in the table shows significant difference at 5% level.

The physical properties of the filling fiber shows that the recycled hollow polyester is lower in all aspects when compared to virgin polyester except in the case of crimp. When compared to Kapok, this fiber is better and hence can be used as pillow filling for the functional pillows.

2.3.2   Comfort Properties of the Filling Fibers (Virgin Hollow Polyester, Recycled Hollow Polyester and 50:50 Blend Virgin Polyester:Recycled Hollow Polyester)

Three different ratios were studied for the filling fibers. The comfort properties of the raw material virgin hollow polyester, recycled hollow polyester and 50:50 Blend of virgin and recycled hollow polyester are given in Fig. 10.

The water vapor permeability of recycled polyester is 0.57% lesser than virgin polyester and when the virgin polyester is blended with recycled polyester in a ratio of 50:50 it is found that there is a 1.79% gain in water vapor permeability compared to that of virgin polyester. The thermal conductivity of recycled polyester is 34% higher than virgin polyester while 50:50 virgin polyester and recycled hollow polyester is 6% higher than that of virgin polyester. The water vapor permeability and thermal

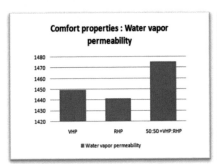 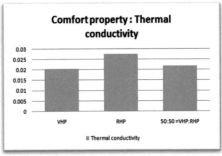

Fig. 10 Comfort properties of the selected filling fibers

conductivity value of fiber need to be high as it helps in the evaporation of sweat and transmits the heat to maintain the temperature. Hence it can be highlighted that 50:50 virgin polyester and recycled hollow polyester prove to be better than virgin polyester and recycled hollow polyester in terms of water vapor permeability and thermal conductivity which are the main features that have a bearing on the comfort of using the pillows.

2.3.3 Test Result of the Yarns Manufactured for Pillow Slip Fabric

Three different combinations of yarns were sourced from Tirupur namely 100% recycled cotton yarn, blend ratios 90:10 and 80:20 recycled cotton: recycled polyester. The properties of the selected yarns are given in Table 8.

The lea strength and single yarn twist of yarn need to be greater as it decides the yarn quality. Lower the count value, thicker the yarn.

From Table 8, it can be understood that the lea Strength of 90:10 Recycled cotton and recycled polyester yarn is 7.5% higher than 100% recycled cotton and lea strength of 80:20 Recycled cotton and recycled polyester yarn is 12.5% higher than 100% Recycled cotton. The Single Yarn Strength of 90:10 Recycled cotton and recycled polyester yarn is 2.47% higher than 100% recycled cotton and single yarn strength of 80:20 Recycled cotton and recycled polyester yarn is 5.13% higher than 100% Recycled cotton. The count of the three samples is almost similar in value.

Hence it can be reported that 80:20 Recycled cotton and recycled polyester yarn is proving to be better than 100% recycled cotton yarn and 90:10 Recycled cotton and recycled polyester yarn in terms of lea strength and single yarn twist. The F value for all parameters shown in Table 8 is significant 5% level.

2.3.4 Test Result of the Pillow Slip Fabric

The physical and mechanical properties of the pillow slip fabrics namely 100% recycled cotton fabric, blend of 90:10 recycled cotton and recycled polyester fabric and blend of 80:20 recycled cotton and recycled polyester fabric, are given in Table 9.

Table 8 Properties of the yarns manufactured for pillow slip fabric

Parameters	Standard 100% RC	90:10 RC:RP	% gain or loss over original	80:20 RC:RP	% gain or loss over original	F value
Lea strength	40	43	7.5	45	12.5	25.21*
Single yarn twist	22.6	23.16	2.477	23.76	5.132	38.64*
Count	19.6	19.6	0	19.5	0.510	9.34*

Table 9 Properties of the developed fabrics for pillow slip

Parameters	Standard 100% RC	90:10 RC:RP	%gain or loss over original	80:20 RC:RP	% gain or loss over original	F value
Thickness (mm)	0.48	0.46	−4.16	0.44	−8.33	90*
GSM (g/m^2)	149.3	134.3	−10.04	124.3	−16.74	712*
Tear strength (gf) warp	953.6	1670.4	75.16	1820.2	90.87	5171*
Tear strength (gf) weft	929.1	1542.4	66.01	1742.2	87.51	230*
Bursting strength (kg/cm^2)	4.9	5.1	4.081	5.3	8.16	25.54*
Crease recovery warp	128°	132°	3.12	133°	3.90	114.96*
Crease recovery weft	132°	137	3.78	140°	6.06	481.44*
Drape co-efficient (%)	85	78	−8.23	69	−18.82	434.25*
Bending modulus (kg/cm^2)	0.40714	0.2531	−37.83	0.2364	−41.93	386.71*

In general, the strength and crease recovery are based on the end use of the product depending on its durability.

From Table 9, it can be understood that the tear strength of 90:10 recycled cotton and recycled polyester fabric increases (75.16%) on the warp and (66.01%) on the weft than 100% recycled cotton. The tear strength of 80:20 recycled cotton and recycled polyester fabric increases (90.87%) on the warp and (87.51%) on the weft than 100% recycled cotton. The Bursting Strength of 90:10 recycled cotton and recycled polyester fabric is 4.08% higher than 100% recycled cotton and that of 80:20 recycled cotton and recycled polyester fabric is 8.16% higher than 100% Recycled cotton. The crease recovery of 90:10 recycled cotton and recycled polyester fabric increases (3.12%) on the warp and (3.78%) on the weft than 100% recycled cotton.

The crease recovery of 80:20 recycled cotton and recycled polyester fabric increases (3.90%) on the warp and increases (6.06%) on the weft. The Drape Coefficient of 90:10 recycled cotton and recycled polyester fabric is 8.23% lesser than 100% recycled cotton and drape coefficient of 80:20 recycled cotton and recycled

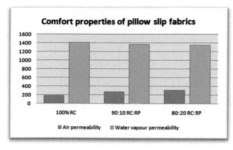

Fig. 11 Comfort properties of pillow slip fabrics

polyester fabric is 18.82% lesser than 100% Recycled cotton. The Bending modulus of 100% recycled cotton is 37.83% higher than 90:10 recycled cotton and recycled polyester fabric and 41.94% higher than 80:20 recycled cotton and recycled polyester fabric. The drape co-efficient and bending modulus also depends upon the thickness of the fabric. The fabric with more thickness will be stiff. Hence the lower the value of drape co-efficient and bending modulus value, the flexibility of fabric will be more.

Hence it can be highlighted that 80:20 Recycled cotton and recycled polyester fabric proved to be better than 100% recycled cotton yarn and 90:10 Recycled cotton and recycled polyester yarn in terms of tear strength, bursting strength and crease recovery. When the polyester content increases in the fabric it is bound to have an impact on the properties of the fabric. The F value for all parameters in Table 9 shows significant difference at 5% level.

The comfort properties of the pillow slip fabrics namely 100% recycled cotton fabric, blend of 90:10 recycled cotton and recycled polyester fabric and blend of 80:20 recycled cotton and recycled polyester fabric, are given in Fig. 11.

The Air permeability of 90:10 and 80:20 recycled cotton and recycled polyester yarn increased by 53.6 and 78.8% respectively when compared to 100% recycled cotton. Water vapor permeability of 90:10 and 80:20 recycled cotton and recycled polyester yarn decreased by 3.1 and 4.3% respectively than 100% recycled cotton fabric. Thermal conductivity of 90:10 and 80:20 recycled cotton and recycled polyester fabric decreased by 5.1 and 10.19% than 100% recycled cotton. Air permeability decides the breathability of a pillow, so the air permeability of the fabric need to high. The water vapor permeability and thermal conductivity value of fabric need to be high as it helps in the evaporation of sweat and transmits the heat to maintain the temperature. Hence it is defined that 100% recycled cotton fabric is proving to be better than 80:20 and 90:10 Recycled cotton and recycled polyester fabric in terms water vapor permeability and thermal conductivity and vice versa in the case of air permeability.

Table 10 Results of the antibacterial finish for the filling fiber and the pillow slip material

Test organism used	Fiber: recycled hollow polyester fiber		Fabric: recycled cotton fabric	
	S. aureus ATCC 6538	*E. coli* ATCC 25922	*S. aureus* ATCC 6538	*E. coli* ATCC 25922
Inoculated bacterial concentration (CFU/ml)	2.1×10^5	2.05×10^5	2.1×10^5	2.05×10^5
Growth value (F) (Criteria = not less than 1.0)	1.07	1.26	1.07	1.26
Bactericidal activity (L) (Criteria = not less than 0)	1.6	3.17	0.82	0.52
Bacteriostatic activity (S) (Criteria = not less than 2.0)	2.9	4.67	2.01	2.02
Measuring method of bacterial concentration	Plate count method	Plate count method	Plate count method	Plate count method

2.3.5   Optimization Results for Functional Pillow Development

The Box-Behnken optimization results showed that the 100% recycled hollow polyester was most suitable for pillow filling, 500 g was the ideal weight of filling material and 100% recycled cotton was recommended for pillow slip cover. Hence these parameters are found to be the optimized conditions for pillow development.

2.3.6   Test Results of Antimicrobial Finish

The antibacterial finish was given to both the filling fiber and the pillow slip fabric to ensure the safety of the material and to satisfy the feeling of the insecurity in the minds of the consumer for using recycled materials. Many images displayed in media show the conditions and the environment where the waste materials are stacked for recycling. The test results of the antibacterial finish for the filling fiber and the pillow slip material are given in Table 10.

The term 'bacteriostatic' means that the agent prevents the growth of bacteria (i.e. it keeps them in the stationary phase of growth) and 'bactericidal' means that antibacterial agent kills bacteria [65]. Recycling sites are always unhygienic, unsafe and unsightly. The recycling waste is piled up providing good environment for the

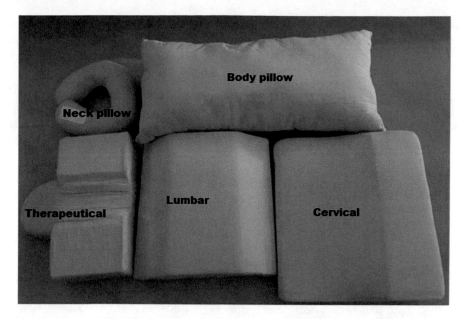

Fig. 12 Images of the functional pillows

formation of bacteria and infection [66]. Apart from regular sequence of processing, antimicrobial finish gives both physiological comfort [67] as well as psychological satisfaction [68]

From the Table 10, the antibacterial treated recycled hollow fibers and recycled cotton fabric show good results to both gram [+ve] and gram [−ve] bacteria with regard to bactericidal and bacteriostatic activity. Pillows are not washed regularly and hence this treatment will stay for a long while and help in the prevention of the growth of bacteria. This treatment is also suitable for the care and maintenance regime (machine wash/sundrying) of the pillows.

2.3.7   Test Results of Functional Pillows

The functional pillows developed are given in Fig. 12. The physical properties of the developed pillows along with compression and recovery are given in Table 11.

Table 11 shows good results for the compression and the recovery properties of all the developed pillows ranging from 1.5 to 7.6% and the recovery is 88–100% when compared to the original weight and thickness. With regard to the comfort properties, the pillows made from compressed fiber showed lower air permeability, water vapor permeability and air permeability than the pillows filled with the hollow recycled polyester. The pressing of the fiber is done to maintain the shape as in lumbar, cervical and therapeutical pillows.

Table 11 Properties of the developed functional pillows

Parameters	Weight (g)	Thickness (in.)	Compression (%)	Recovery (%)	Thermal conductivity	Water vapor permeability	Air permeability
Lumbar pillow	845	5.3	5.6	100	0.0461	5735.60	65.8
Neck pillow	230	3.25	7.6	88	0.0552	6774.72	75.5
Cervical pillow	995	5.75	1.7	100	0.0331	5821.63	68.4
		3.25	2.8	100	0.0541	6774.72	70.5
Therapeutical pillow	890	6	1.6	100	0.0300	5254.83	64.5
		4	2.5	100	0.0459	6335.88	70.3
Body pillow	1100	7.25	4.1	90	0.0556	6665.12	72.8

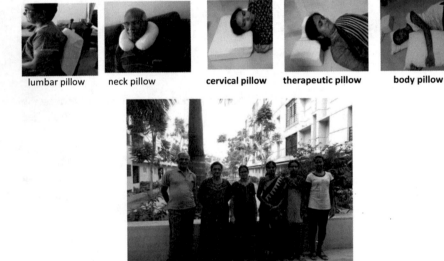

lumbar pillow   neck pillow   cervical pillow   therapeutic pillow   body pillow

Fig. 13 The respondents who have used the functional pillows

2.3.8   Feedback Analysis for Functional Pillows

Respondents were given the functional pillows for feedback. Most of them felt that the pillows were user friendly, very suitable to the body shape, sleep inducing, odor free, feeling of safety due to anti bacterial treatment, evaporates sweat easily and caused no allergic reactions. Figure 13 shows some of the respondents who used the functional pillows for a period of one month.

2.3.9   Cost Savings Analysis

The cost estimation for developing the functional pillows has been given in Table 12. This estimation is compared with functional pillows developed with Virgin hollow polyester and virgin pillow slip cover material. Apart from this estimation there are other benefits in terms of energy, CO_2 equivalents, transportation etc. which has not been taken up in this study. Literature and calculators are available to estimate the outcome of using recycled materials. The parameters for estimating the cost is based on the amount spent by the researchers for each and every procurement and processing; when product development is taken on large scale commercial production cost ratios, the benefits may be still larger in the case of recycled materials.

Table 12 highlights the savings in terms of unit price for each design developed. Virgin hollow polyester fibre and 100% Virgin cotton fabric is compared with recycled hollow polyester fibre and 100% recycled cotton for all five functional pillows. In the case of raw material there is a meagre reduction in cost when compared to the 'all virgin materials'. When the percentage of recycled material increases the savings is higher. In the case of unit price comparison of functional pillows there is a savings in cost in the case of recycled products ranging from 0.12 to 1.4%.

For the sake of cost estimation, it has been assumed that the annual turnover of a small scale industry manufacturing pillows, not functional pillows, is Rs. 20,00,000 (Rupees twenty lakhs) and the standard rate per pillow as Rs. 500 (Rupees five hundred) irrespective of the pillow design. Table 13 shows that the savings based on annual turnover ranges from Rs. 2502 to 27,361/-. Similarly the savings based on annual production ranges from Rs. 6005 to 65,667/-. This indicates that the use of recycled materials is economical and is beneficial both in terms of raw material reduction and investment on raw material. The savings will increase if the actual functional pillow cost is compared in the case of virgin material instead of a standard rate of Rs. 500/- per pillow. Apart from this savings estimate there are other savings in terms of raw material usage and environmental impact.

2.3.10   Conclusion

Sustainability is the key aspect in today's manufacturing and the greatest tool to maintain sustainability is recycling. Consumers today look out for products with recycled labels. The major concern is our environment and scientists look out for methods to reduce the pollution levels in the society. This project is aimed at developing functional pillow using recycled hollow polyester fiber from PET bottle waste and recycled cotton fabric developed from apparel industry cut waste. Antimicrobial coating is given to both fiber and fabric to give the consumer confidence and safety feeling to use recycled products.

From the test results it can be concluded that the therapeutic pillow gave the best results in terms of physical, mechanical and comfort properties. The feedback given by the respondents also show acceptance of the functional pillows with recycled materials. In the case of saving analysis all the five type of pillows show the same

Table 12 Cost estimation of functional pillows[a]

Particulars	Lumbar pillow		Neck pillow		Cervical pillow		Therapeutical pillow		Body pillow	
	RHP (750 g)	VHP (750 g)	RHP (160 g)	VHP (160 g)	RHP (875 g)	VHP (875 g)	RHP (771 g)	VHP (771 g)	RHP (977 g)	VHP (977 g)
Fiber requirement	150	187.5	32	40	175	218.75	154.2	192.75	195.4	244.25
Fiber cost										
Cleaning charges	45	0	9.6	0	52.5	0	46.2	0	58.6	0
Antimicrobial finish	183.75	183.75	39.2	39.2	214.37	214.37	188.89	188.89	239.36	239.36
Thermal bonded fiber	200	200	0	0	200	200	200	200	0	0
Fabric	RC (60 cm)	VC (60 cm)	RC (45 cm)	VC (45 cm)	RC (60 cm)	VC (60 cm)	RC (75 cm)	VC (75 cm)	RC (100 cm)	VC (100 cm)
yarn cost	19.99	21.49	14.99	16.11	19.99	21.49	24.99	26.86	33.32	35.81
Fabric cost	138	147	103.5	110.25	138	147	172.5	183.75	230	245
Chemical processing	177	177	132.75	132.75	177	177	221.25	221.25	295	295
Stitching charge	420	420	120	120	420	420	720	720	100	100
Total	1333.74	1336.74	452.04	458.31	1396.86	1398.61	1728.03	1733.5	1151.68	1159.42
Savings % per pillow	**0.22**		**1.36**		**0.12**		**0.31**		**0.66**	

[a] All calculations subject to change

Table 13 Saving analysis based on turnover and annual production

Particulars	Standard virgin pillow cost	Recycled pillow cost	Gain or loss % over standard pillow cost	[b]Savings based on turnover	[a]Savings based on annual production
Lumbar pillow	1333.74	1336.74	0.22	4488.53	10,772.48
Neck pillow	452.04	458.31	1.36	27,361.40	65,667.34
Cervical pillow	1396.86	1398.61	0.12	2502.48	6005.96
Therapeutical pillow	1728.03	1733.5	0.31	6310.93	15,146.24
Body pillow	1151.68	1159.42	0.66	13,351.5	32,043.61

All calculations subject to change
[a]All calculations are based on commercial unit price of Rs. 500 per pillow: annual production of 48,000 pillows
[b]All calculations based on annual turnover of Rs. 20,00,000/-

trend—lower cost for pillows using recycled materials. Conversion of waste material to enter the raw material stream in a production system will reduce the use of virgin material, thereby saving the virgin raw material usage and reduction in environmental pollution. Hence this process can be incorporated into a production system to make it sustainable and eco-friendly.

3 Roadmap for Sustainability by Use of Recycled Materials

The main aspect of sustainability by the use of recycled materials is environmental improvements. Virgin material substitute is recycled material and there are many strategies that can be used to attain this objective. The strategies are termed as *Replacement, Reduction, Transformation and Reorganization* to help in ease of adoption as shown in Fig. 14. A report states that the use of 2.9 billion PET, which have been diverted from landfill, produced 47,407 m and saves 2789 million MJ fossil fuel energy and 89,699 tons of CO_2 eq [69]. The role of stake holders in promoting the role of recycled materials highlights the importance of using recycled materials not only for savings in virgin materials but also in terms of environmental impact. Changes in attitude of consumers, awareness in doing business calculating the environmental impact and fine tuning manufacturing towards latest sustainable technologies are of utmost importance. A textile system with a higher rate of clothing utilization, reduction of waste in manufacturing and improved recycling will foster lower environmental impacts and help to move towards sustainability. In the case of recycled polyester, bountiful benefits results as PET bottles do not reach land-

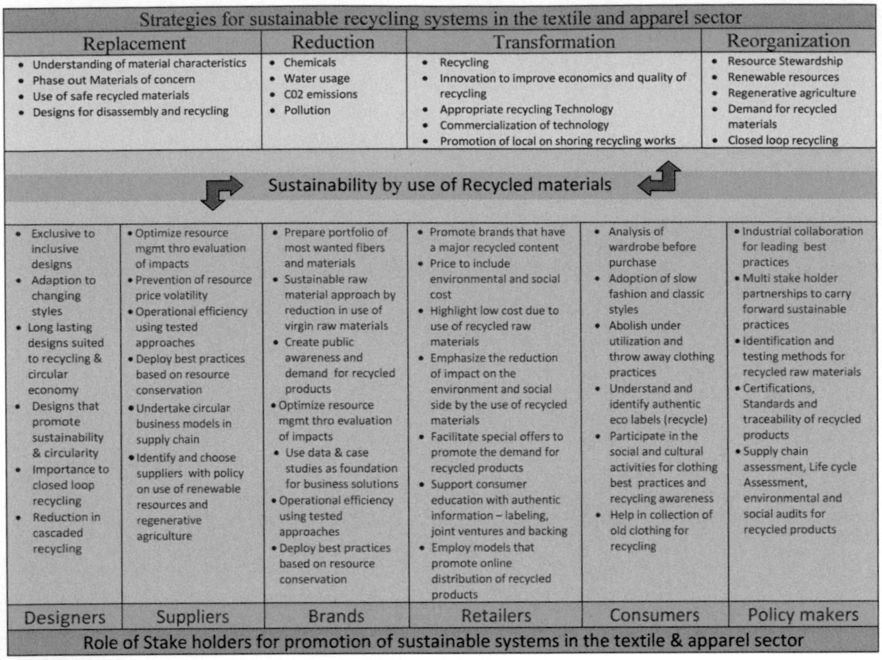

Strategies for sustainable recycling systems in the textile and apparel sector			
Replacement	Reduction	Transformation	Reorganization
• Understanding of material characteristics • Phase out Materials of concern • Use of safe recycled materials • Designs for disassembly and recycling	• Chemicals • Water usage • CO2 emissions • Pollution	• Recycling • Innovation to improve economics and quality of recycling • Appropriate recycling Technology • Commercialization of technology • Promotion of local on shoring recycling works	• Resource Stewardship • Renewable resources • Regenerative agriculture • Demand for recycled materials • Closed loop recycling

Sustainability by use of Recycled materials

Designers	Suppliers	Brands	Retailers	Consumers	Policy makers
• Exclusive to inclusive designs • Adaption to changing styles • Long lasting designs suited to recycling & circular economy • Designs that promote sustainability & circularity • Importance to closed loop recycling • Reduction in cascaded recycling	• Optimize resource mgmt thro evaluation of impacts • Prevention of resource price volatility • Operational efficiency using tested approaches • Deploy best practices based on resource conservation • Undertake circular business models in supply chain • Identify and choose suppliers with policy on use of renewable resources and regenerative agriculture	• Prepare portfolio of most wanted fibers and materials • Sustainable raw material approach by reduction in use of virgin raw materials • Create public awareness and demand for recycled products • Optimize resource mgmt thro evaluation of impacts • Use data & case studies as foundation for business solutions • Operational efficiency using tested approaches • Deploy best practices based on resource conservation	• Promote brands that have a major recycled content • Price to include environmental and social cost • Highlight low cost due to use of recycled raw materials • Emphasize the reduction of impact on the environment and social side by the use of recycled materials • Facilitate special offers to promote the demand for recycled products • Support consumer education with authentic information – labeling, joint ventures and backing • Employ models that promote online distribution of recycled products	• Analysis of wardrobe before purchase • Adoption of slow fashion and classic styles • Abolish under utilization and throw away clothing practices • Understand and identify authentic eco labels (recycle) • Participate in the social and cultural activities for clothing best practices and recycling awareness • Help in collection of old clothing for recycling	• Industrial collaboration for leading best practices • Multi stake holder partnerships to carry forward sustainable practices • Identification and testing methods for recycled raw materials • Certifications, Standards and traceability of recycled products • Supply chain assessment, Life cycle Assessment, environmental and social audits for recycled products

Role of Stake holders for promotion of sustainable systems in the textile & apparel sector

Fig. 14 Roadmap for sustainability by use of recycled materials

fills but become feedstock for the next cycle of textile manufacturing leading the environmental benefits and resource.

A report by a research group in Austria states that the carbon footprint for RPET shows savings of around 79%. The operational procedure for the production of RPET followed by the organization showed a carbon footprint of 0.45 kg CO_2 eq per kg of RPET while virgin PET had a carbon equivalent of 2.15 per kilogram. The reduction of 1.7 kg accounts to 79% which can be used to power a 13 W bulb continuously for twenty days as per the Austrian Standards [70]. The calculations were based on ISO 14044 and the electricity and gas consumption showed the mass and energy balance for all the different stages of production.

4 Concluding Remarks

rPET is a versatile material that can replace virgin raw material for different applications. When chemical and thermal recycling was initially carried out the recycled polyester showed characteristics of strength reduction which proved that it could not be recycled continuously to be used in the production cycle again and again. The recycled material could be used to produce lower grade goods in terms of quality. In recent times many researchers have proved through various studies and publications that polyester could be recycled again and again to move into the production cycle equaling virgin polyester. In China, under the Recycling Subsidy Policy, the government gives subsidy to recyclers or consumers based on the collection and used products and thereby develop the recycling and reuse industry [71]. Similarly all nations will work towards sustainability through tools like reuse and recycling.

Literature on Recycling exist based on assumptions like the gaming theory or optimization modeling without concern about the interactions between the stake holders like government, manufacturers, recyclers and consumers. Currently the System Dynamics (SD) simulation method is applied to analyze business policies and problems in the operations of recycling [72–75] This system works on the production techniques in remanufacturing, management of reverse logistics and the design features of the closed loop supply chain network, technological innovations in recycling and use of non-renewable resources. The MADE-BY fiber benchmark has classified all recycled fibers into grades A to E based on their environmental performance in terms of GHG emissions, human toxicity, eco toxicity, energy, water and landuse. Recycled cotton and mechanically recycled polyester score in the Class A; Chemically recycled polyester comes under Class B while Virgin polyester scores the Class D and conventional cotton comes under Class E [76].

Awareness in having a recycled component in the product has spread across the globe. Many standards like Control Union Global Recycle Standards, REMO, Recycled Claim Standard and Cradle to Cradle certificate are some of the tools to label the recycled products and get consumer's choice. The supply chain is also widening by providing recycled raw materials to manufacturers and designers namely Recycled polyester fibers from PET bottles Eco-fi, Unifi, RAW Bionic yarn from reclaimed sea plastic waste etc. Opportunities are expanding and it is the responsibility of designers, manufacturers and the governments to take up the task of promoting and evaluating recycling of waste and rPET has a major role to play in safeguarding the environment.

References

1. Lowe ED (2019) Polyester. In: Beauty & fashion. Available via https://fashion-history. lovetoknow.com/fabrics-fibers/polyester. Accessed 17 Mar 2019
2. Boekhodd K (1996) History. In: Polyester. Available via http://schwartz.eng.auburn.edu/ polyester/history.html. Accessed 17 Mar 2019
3. Hollie PG (1983) Bid for better polyester image. The New York Times. https://www.nytimes. com/1983/04/04/business/bid-for-better-polyester-image.html Accessed 17 Mar 2019
4. Krapp KM (2019) Polyester. In: How products are made. Available via http://www.madehow. com/Volume-2/Polyester.html. Accessed 18 Mar 2019
5. What is Polyester (2015) History of polyester. Available via http://www.whatispolyester.com/ history.html. Accessed 18 Mar 2019
6. Statistica (2019) Production of polyester fibers worldwide from 1975–2017 (in 1,000 metric tons). In: The Statistics Portal https://www.statista.com/statistics/912301/polyester-fiber-production-worldwide/. Accessed 18 Mar 2019

7. IHS Markit (2018) Polyester fibers. In: Chemical economics handbook. Available via https://ihsmarkit.com/products/polyester-fibers-chemical-economics-handbook.html. Accessed 18 Mar 2019
8. The House of Pillows (2019) The dangers of polyester and how it hurts our environment and wild life. Available via https://www.thehouseofpillows.eu/polyester-production-blog/. Accessed 20 Mar 2019
9. Aqua-calc (2019) Cubic meters to liters. Available via https://www.aqua-calc.com/convert/volume/cubic-meter-to-liter. Accessed 20 Mar 2019
10. Kalliala EM, Nousiainen P (1999) Environ mental profile of cotton and polyester-cotton fabrics. AUTEX RES J 1:8–20. http://www.autexrj.com/cms/zalaczone_pliki/2b.pdf
11. Gerngross TU, Slater SC (2000) How green are green plastics? Sci Am 2000:36–41
12. Akiyama M, Tsuge T, Doi Y (2003) Environmental life cycle comparison of polyhydroxyalkanoates produced from renewable carbon resources by bacterial fermentation. Polym Degrad Stabil 80:183–194
13. Hatch KL (1984) Chemicals and textiles. Part 1: Dermatological problems related to fiber content and dyes. TRJ 54:664–682
14. Lewin M, Pearce EM (1998) Handbook of fiber chemistry, 2nd edn. Marcel Dekker, New York
15. Sherrington C (2016) Plastics in the marine environment. In: Eunomia. Available via https://www.eunomia.co.uk/reports-tools/plastics-in-the-marine-environment/. Accessed 21 Mar 2019
16. Pennsylvania State University (2019) Tiny fibers create unseen plastic pollution. In: Phys.org. Available via https://phys.org/news/2019-02-tiny-fibers-unseen-plastic-pollution.html. Accessed 21 Mar 2019
17. Messinger L (2016) How our clothes are poisoning our oceans and food supply. Available via https://www.theguardian.com/environment/2016/jun/20/microfibers-plastic-pollution-oceans-patagonia-synthetic-clothes-microbeads. Accessed 23 Mar 2019
18. University of Barcelona (2018) Presence of textile microfibers from washing machines in marine floors. In: Science Daily. Available via https://www.sciencedaily.com/releases/2018/11/181115115355.htm. Accessed 20 Mar 2019
19. Bruce N, Hartline N, Karba S, Ruff B, Sonar S, Holden P (2016) Microfiber pollution and the apparel industry. Available via http://www.esm.ucsb.edu/research/2016Group_Projects/documents/PataPlastFinalReport.pdf. Accessed 20 Mar 2019
20. Browne MA, Crump P, Niven SJ, Teuten EL, Tonkin A, Galloway T, Thompson RC (2011) Accumulations of microplastic on shorelines worldwide: sources and sinks. Environ Sci Technol. https://doi.org/10.1021/es201811s. Available via https://imedea.uib-csic.es/master/cambioglobal/Modulo_III_cod101608/tema%2011-invasoras%202013-2014/plastics/Browne_2011-EST-Accumulation_of_microplastics-worldwide-sources-sinks.pdf. Accessed 21 Mar 2019
21. Vidal AS, Thompson RC, Canals M, Haan WP (2018) The imprint of microfibers in southern Europe. In: PLoS ONE. https://doi.org/10.1371/journal.pone.0207033. Available via https://journals.plos.org/plosone/article?id=10.1371/journal.pone.0207033. Accessed 21 Mar 2019
22. Henry B, Laitala K, Klepp I G (2019) Microfibers from apparel and home textiles: Prospects for including microplastics in environmental sustainability assessment. Sci Total Environ 652:438–494. https://www.sciencedirect.com/science/article/pii/S004896971834049X
23. Youngsteadt E (2011) Laundry lint pollutes the World's Oceans. In: Science. Available via https://www.sciencemag.org/news/2011/10/laundry-lint-pollutes-worlds-oceans. Accessed 22 Mar 2019
24. Cumarsaide R, hAeraide Gniomhaithe ar son na & Comhshaoil (2019) Waste hierarchy. Available via https://www.dccae.gov.ie/en-ie/environment/topics/waste/waste-management-and-policy/Pages/Waste-Hierarchy.aspx. Accessed 22 Mar 2019
25. DEFRA (2011) Waste hierarchy. Available via https://assets.publishing.service.gov.uk/government/uploads/system/uploads/attachment_data/file/69403/pb13530-waste-hierarchy-guidance.pdf. Accessed 22 Mar 2019

26. ISO (2019a) ISO 15270:2008 Plastics—guidelines for the recovery and recycling of plastics waste. Available via https://www.iso.org/standard/45089.html. Accessed 25 Mar 2019

27. ISO (2019b) ISO 14001:2015 Environment management systems—requirements with guidance for use. Available via https://www.iso.org/standard/60857.html. Accessed 24 Mar 2019

28. Last Bottle Clothing (2019) The ecological effects of cotton vs recycled polyester. Available via https://lastbottleclothing.com/the-ecological-effects-of-cotton-vs-recycled-polyester/. Accessed 21 Mar 2019

29. Tapia-Picazo JC, Luna-Bárcenas JG, García-Chávez A, Gonzalez-Nuñez R, Bonilla-Petriciolet A, Alvarez-Castillo A (2014) Polyester fiber production using virgin and recycled PET. Fiber Polym 15:547–552

30. Textile Mates (2017) Comparative study of virgin and recycled polyester. Available via https://www.textilemates.com/virgin-recycled-polyester/. Accessed 17 Mar 2019

31. He SS, Wei MY, Liu MH, Xue WL (2015) Characterization of virgin and recycled poly (ethylene terephthalate) (PET) fibers. JTI 106:800–806

32. Yuksekkaya ME, Celep G, Dogan G, Tercan M, Urhan B (2016) A comparative study of physical properties of yarns and fabrics produced from virgin and recycled fibers. JEFF 11:68–76

33. Mancini SD, Zanin M (1999) Recyclability of PET from virgin resin. Mater Res 2:33–38

34. Wróbel G, Bagsik R (2010) The importance of PET filtration for the possibility of material recycling. J Achiev Mater Manuf Eng 43:178–191

35. Frounchi M, Mehrabzadeh M, Ghiaee R (1997) Studies on recycling of polyethylene terephthalate beverage bottles. Iran Polym J 6:269–272

36. Spinacé MS, De Paoli MA (2001) Characterization of poly (ethylene terephtalate) after multiple processing cycles. J Appl Polym Sci 80:20–25

37. Elven MV (2018) How sustainable is recycled polyester. Available via https://fashionunited.uk/news/fashion/how-sustainable-is-recycled-polyester/2018111540000. Accessed 21 Mar 2019

38. Aizenshtein EM (2016) Bottle wastes to textile yarns. Fibre Chem 47:343–347. https://doi.org/10.1007/s10692-016-9691-8

39. Plastics Europe (2015) Plastics—the facts 2014/2015 an analysis of European plastics production, demand and waste data. Available via https://www.plasticseurope.org/application/files/5515/1689/9220/2014plastics_the_facts_PubFeb2015.pdf. Accessed 20 Mar 2019

40. Kuczenski B and Geyer R (2014) Life cycle assessment of polyethylene terephthalate (pet) beverage bottles consumed in the State of California: Contractor's report. Sacramento, CA

41. Telli A, Özdil N, Babaarslan O (2012) Usage of PET bottle wastes in textile industry and contribution to sustainability. J Text Eng 19:86

42. Resource Center (2019) What is closed-loop recycling? Available via https://www.buschsystems.com/resource-center/knowledgeBase/glossary/what-is-closed-loop-recycling. Accessed 21 Mar 2019

43. Sustainability Dictionary (2019) Available via https://sustainabilitydictionary.com/2005/12/03/open-loop-recycling/. Accessed 21 Mar 2019

44. Ross C (2015) What's the deal with recycled polyester. Available via https://theswatchbook.offsetwarehouse.com/2015/01/29/what-is-recycled-polyester/. Accessed 21 Mar 2019

45. Radhakrishnan S (2017) Denim recycling. Textiles and clothing sustainability: recycled and upcycled textiles and fashion. Springer, Singapore, pp 79–125. https://doi.org/10.1007/978-981-10-2146-6

46. Carbios (2019) CARBIOS: using enzymes to "biorecycle" PET. Available via http://greenblueorg.s3.amazonaws.com/smm/wp-content/uploads/2017/10/Carbios.pdf. Accessed 21 Mar 2019

47. Gr3n (2019) Gr3n: a new approach to PET chemical recycling. Available via http://greenblueorg.s3.amazonaws.com/smm/wp-content/uploads/2017/10/Gr3n.pdf. Accessed 21 Mar 2019

48. Loop Industries(2019) Loop Industries: recycling PET waste into high-quality resin. Available via http://greenblueorg.s3.amazonaws.com/smm/wp-content/uploads/2017/10/Loop-Industries.pdf. Accessed 21 Mar 2019

49. Resinate Materials Group (2019) Resinate Materials Group: Turning PET waste into high performance polyester polyols. Available via http://greenblueorg.s3.amazonaws.com/smm/wp-content/uploads/2017/11/Resinate_-1.pdf. Accessed 21 Mar 2019

50. Koh H (2017) A way to repeatedly recycle polyester has just been discovered. Available via https://www.eco-business.com/news/a-way-to-repeatedly-recycle-polyester-has-just-been-discovered/. Accessed 22 Mar 2019

51. Textile Environmental Design (2019) Polyester recycling. Available via http://www.tedresearch.net/media/files/Polyester_Recycling.pdf. Accessed 21 Mar 2019

52. GFN (2019a) FAQs general. Available via https://www.footprintnetwork.org/faq/. Accessed 25 Mar 2019

53. GFN (2019b) Ecological footprint. Available via https://www.footprintnetwork.org/our-work/ecological-footprint/. Accessed 25 Mar 2019

54. Zamani B, Svanström M, Peters G, Rydberg T (2015) A carbon footprint of textile recycling: a case study in Sweden. J Ind Ecol 19:676–687

55. Woolridge AC, Ward GD, Phillips PS, Collins M, Gandy S (2006) Life cycle assessment for reuse/recycling of donated waste textiles compared to use of virgin material: an UK energy saving perspective. Resour Conserv Recycl 46(1):94–103

56. Kamath MG, Bhat GS (2008) Specialty fibers from polyesters and polyamides. In: Polyesters and polyamides. Woodhead Publishing, pp 203–218

57. Khoddami A, Carr CM, Gong RH (2009) Effect of hollow polyester fibres on mechanical properties of knitted wool/polyester fabrics. Fibers Polym 10:452–460

58. Snyder AC (1994) U.S. Patent No. 5,344,707. U.S. Patent and Trademark Office, Washington, DC. Available via https://patents.google.com/patent/US5344707A/en. Accessed 1 Apr 2019

59. Neimark J (2005) Are your pillows hazardous to your health? Available at https://www.prohealth.com/library/are-your-pillows-hazardous-to-your-health-22877 (21 August 2018)

60. SandraLaville S, Taylor M (2017) A million bottles a minute: world's plastic binge 'as dangerous as climate change'. Available via https://www.theguardian.com/environment/2017/jun/28/a-million-a-minute-worlds-plastic-bottle-binge-as-dangerous-as-climate-change. Accessed 21 August 2018

61. Chaterjee B (2017) India recycles 90% of its PET waste, outperforms Japan, Europe and US: study. Available via https://www.hindustantimes.com/mumbai-news/india-recycles-90-of-its-pet-waste-outperforms-japan-europe-and-us-study/story-yqphS1w2GdlwMYPgPtyb2L.html. Accessed 20 Aug 2018

62. Chaterjee S (2017a) Do indian government provide any subsidy for the plastic recycling plant? Available via https://www.quora.com/Do-indian-government-provide-any-subsidy-for-the-plastic-recycling-plant. Accessed 19 Aug 2018

63. Au Lit Fine Linens (2018) Perfect pillow sizes: standard, queen, or king? Available via https://www.aulitfinelinens.com/blogs/betweenthesheets/72346309-perfect-pillow-sizes-standard-queen-or-king. Accessed 19 Aug 2018

64. Graphic Pad (2017) One way ANOVA. Available via https://www.graphpad.com/guides/prism/7/statistics/f_ratio_and_anova_table_(one-way_anova).htm?toc=0&printWindow. Accessed 25 Mar 2019

65. Pankey GA, Sabath LD (2004) Clinical relevance of bacteriostatic versus bactericidal mechanisms of action in the treatment of Gram-positive bacterial infections. Clin Infect Dis 38(6):864–870

66. CEF (2019) Advantages and disadvantages of recycling. https://www.conserve-energy-future.com/advantages-and-disadvantages-of-recycling.php. Accessed 25 Mar 2019

67. Hui X, Zhu H, Sun G (2016) Antimicrobial textiles for treating skin infections and atopic dermatitis. In: Antimicrobial textiles. Woodhead Publishing, pp 287–303

68. Cloud RM, Cao W, Song G (2013) Chapter 11. Functional finishes to improve the comfort and protection of apparel. In: Advances in the dyeing and finishing of technical textiles. Elsevier Inc.

69. Textile Exchange (2018) Threading the needle https://tax.kpmg.us/content/dam/tax/en/taxwatch/pdfs/2018/kpmg-threading-needle-report.pdf. Accessed 1 May 2019

70. Alpha (2017) Study confirms the excellent carbon footprint of recycled PET. https://blog.alpla.com/en/press-release/newsroom/study-confirms-excellent-carbon-footprint-recycled-pet/08-17. Accessed 1 May 2019
71. Chang X, Fan J, Zhao Y, Wu J (2016) Impact of China's recycling subsidy policy in the product life cycle. Sustainability 8(8):781
72. Poles R (2013) System dynamics modelling of a production and inventory system for remanufacturing to evaluate system improvement strategies. Int J Prod Econ 144(1):189–199
73. Das D, Dutta P (2013) A system dynamics framework for integrated reverse supply chain with three way recovery and product exchange policy. Comput Ind Eng 66(4):720–733
74. Sterman JD (2000) Business dynamics: systems thinking and modeling for a complex world (No. HD30. 2 S7835 2000)
75. Georgiadis P, Vlachos D (2004) The effect of environmental parameters on product recovery. Eur J Oper Res 157(2):449–464
76. Madeby (2018) Made-by environmental benchmark for fibres. Available via https://www.google.com/search?q=Fiber+Benchmark+(MADE-BY)&oq=Fiber+Benchmark+(MADE-BY)&aqs=chrome..69i57j0.2806j0j8&sourceid=chrome&ie=UTF-8. Accessed 4 Apr 2019

Case Studies on Recycled Polyesters and Different Applications

P. Senthil Kumar and P. R. Yaashikaa

Abstract The developing textile sector confronts difficulties to recycle and reuse their material wastes into applicable industrial products. Recycling is the most suitable way to deal with diminishing the solid waste. Polyester is the world's common recyclable polymer. Recycled polyester utilizes PET as the crude material. The enthusiasm for polyethylene terephthalate (PET) recycling gained attention in recent years; however it has been practised consistently in the past years. Recycled polyesters are used in food contact bottles, sheeting, films, fibres, etc. Polyester, an engineered fibre produced using oil, a non-renewable asset, is generally known for its eco-friendly impacts at the time of its production processes. Utilizations of recycled polyester fibres in the manufacturing of clothing are getting sociable nowadays. Drink bottles delivered from reused PET are on the ascent. Recycled polyester generated from the post-consumer waste, for example, PET bottles are observed to be ecologically advantageous contrasted with origin polyester. The techniques used for recycling can be classified as primary, secondary, tertiary or quaternary methodologies. The two currently used techniques are mechanical and chemical recycling of polyester. Recycling of polyester is advancing to fulfil the necessities of the requirements. Various utilizations are being processed from recycled polyester. Acceptance by consumer remains a major benchmark for sustainable recycling polyester. The strategy utilized in this examination is from contextual analyses. The contextual analyses assessed application, recycling, process procedures and overall polyester recycling. The discoveries feature powerful headway and perspective to be connected in the proposed recycling process. The point of this investigation is to survey the present polyester recycling system so as to advance and assemble a manageable polyester recycling framework.

Keywords Polyester · Recycling · Methods · Applications · Case studies · Waste management

P. Senthil Kumar (✉) · P. R. Yaashikaa
Department of Chemical Engineering, SSN College of Engineering, Chennai 603110, India
e-mail: senthilkumarp@ssn.edu.in

P. Senthil Kumar
SSN-Centre for Radiation, Environmental Science and Technology (SSN-CREST), SSN College of Engineering, Chennai 603110, India

© Springer Nature Singapore Pte Ltd. 2020
S. S. Muthu (ed.), *Environmental Footprints of Recycled Polyester*, Textile Science and Clothing Technology, https://doi.org/10.1007/978-981-13-9578-9_4

1 Introduction

Polyester is a class of polymers that contain the ester useful gathering in their primary chain. A build up response that is fundamentally the same as the response used to make polyamide or nylons frames polyesters [1]. Polyesters and polycarbonates are plastics delivered by the polycondensation of dialcohols with diprotonic acids, a dicarboxylic either corrosive or carbonic corrosive as its corrosive dichloride phosgene [2]. The subsequent ester joins are touchy to hydrolysis. Particularly amid handling, it is important to work with dry polymers. Indeed, even the nearness of little amounts of water causes the deficiency of the chain length at the handling temperature of more than 250 °C, bringing about an undesirable difference in properties. Nevertheless, this impact can be utilized after use to debase polyesters and polycarbonates to get monomers or other profitable items. Today, the attention is on the reuse of waste materials. Land filling is restricted by space and environmental concerns. Both physical and concoction recycling are potential outcomes for these materials. Nonetheless, the physical recycling of polymers is constantly associated with the loss of properties (down cycling), because of the utilization of materials with various chain lengths, distinctive added substances and an alternate history. Synthetic recycling (feedstock recycling) must be created for each material as indicated by its structure and reactivity. Procedures reasonable for polymers as polyolefins, polystyrene or poly (methyl methacrylate) are normally not proper for polycondensates [3]. Polymers can be effectively corrupted by pyrolysis bringing about monomers, for example, styrene, methyl methacrylate, ethylene or propylene. Waxes and hydrocarbon oils can likewise be acquired relying upon the response temperature. Arbitrary scission of the polymer spine discharges essentially ethylene subordinates from polymerizates. Blended plastics convey oils with qualities like raw petroleum. Nevertheless, polycondensates will in general carbonize amid the corruption. The concoction structure of the monomer is pulverized by the corruption of the macromolecule discharging the practical gathering, for example CO_2 and CO because of polyesters [2] and polycarbonates [3] or HCN because of polyamides [4]. The high proportion of sweet-smelling structures incorporated into the structure of many polycondensates advances the development of carboneous deposits. Solvolysis is an increasingly proficient route for the corruption of polycondensates. At temperatures underneath pyrolysis temperature, a nucleophilic operator stops the material. The first monomer or a monomer subsidiary is recuperated. The nonattendance of radical responses underneath the pyrolysis temperatures counteracts the arrangement of carboneous build-up. The roadmap of the present chapter were shown in Fig. 1.

2 Polyesters

As a particular material, it most regularly alludes to a sort called polyethylene terephthalate (PET). Polyesters incorporate normally happening synthetic compounds, for example, in the cutin of plant fingernail skin, just as synthetics, for example,

Fig. 1 Roadmap of recycling polyester

polybutyrate. Characteristic polyesters and a couple of manufactured ones are biodegradable, yet most engineered polyesters are definitely not. The material is utilized widely in garments. Polyester strands are now and then spun together with characteristic filaments to create a fabric with mixed properties. Cotton-polyester mixes (polycotton) can be solid, wrinkle and tear-safe, and diminish contracting [4]. Manufactured filaments utilizing polyester have high water, wind and ecological opposition contrasted with plant-inferred strands. They are less fireproof and can soften when touched off. Polyester mixes have been renamed in order to recommend their closeness or even prevalence over regular filaments. Contingent upon the synthetic structure, polyester can be a thermoplastic or thermoset. There are likewise polyester gums restored by hardeners; in any case, the most well-known polyesters are thermoplastics. Instances of thermoset polyesters incorporate a portion of the Desmophen brand from Bayer. The OH bunch is responded with an Isocyanate practical compound in a 2-segment framework delivering coatings, which may alternatively be pigmented. Polyesters as thermoplastics may change shape after the utilization of warmth. While flammable at high temperatures, polyesters will in general psychologist far from flares and self-douse upon start. Polyester filaments have high diligence and E-modulus just as low water assimilation and insignificant shrinkage in correlation with other mechanical strands. Unsaturated polyesters (UPR) are thermosetting saps. They are utilized in the fluid state as throwing materials, in sheet shaping mixes, as fiberglass covering gums and in non-metallic auto-body fillers [5]. They are additionally utilized as the thermoset polymer framework in pre-pregs. Fiberglass-fortified unsaturated polyesters find wide application in groups of yachts and as body portions of vehicles.

2.1 Applications of Polyesters

Textures woven or sewed from polyester string or yarn are utilized broadly in clothing, home goods, from shirts and jeans to coats and caps, bed sheets, covers, upholstered furniture, and PC mouse mats. Mechanical polyester strands, yarns and ropes are utilized in vehicle tire fortifications, textures for transport lines, seat straps, covered textures and plastic fortifications with high-vitality retention. Polyester fibre is utilized as padding and protecting material in cushions, sofas and upholstery cushioning [6]. Polyester textures are exceedingly recolour safe—indeed, the main class of colours, which can be utilized to adjust the shade of polyester texture, are what are known as scatter colours. Polyesters are likewise used to make bottles, films, canvas, kayaks, fluid precious stone presentations, 3D images, channels, dielectric film for capacitors, film protection for wire and protecting tapes. Polyesters are generally utilized as a completion on top-notch wood items, for example, guitars, pianos and vehicle/yacht insides. Thixotropic properties of shower relevant polyesters make them perfect for use on open-grain timbers, as they can rapidly fill wood grain, with a high-form film thickness per coat. Restored polyesters can be sanded and cleaned to a polished, tough completion.

2.2 Recycling Polyester

Polyester recycling dependably implies that the polymeric material was at that point changed, some way or another, into a semi-completed item or a final result and that it is delivered amid the preparing stage or in the wake of being utilized, or that it is accumulated as a material so it very well may be driven back to the creation of polymeric semi-completed items or polymeric items. When recycling, two different ways for the most part must be separated. The concoction change into introductory crude materials, the polymer structure being crushed, and the physical treatment, the underlying polymer properties being kept up or reconstituted. With polyester, crude material recycling prompts terephthalic corrosive or dimethyl terephthalate and ethylene glycol; with polyamide 6, it prompts caprolactam. These recycling forms just turn out to be cost-productive with huge lines >50,000 tons/year are yet a special case today. Such lines must be seen, heretofore, in the creation focuses of enormous makers like Eastman, DuPont or KOSA. Direct course in a polymeric state is today worked in the most various variations; the procedures utilized for this are regular of the medium-sized industry, and cost proficiency would already be able to be accomplished with line sizes of approx. 10,000 tons/year. For this situation, all types of reused material criticism into the material dissemination are conceivable, the recycling procedure PET-bottle back to the container, for instance, being a unique case. In the future, the differing recycling forms—specifically with respect to the cleaning of remote substances—are being checked, the filtration of polymeric melts and middle of the road items having a specific influence. Other than the synthetic stores and debasement items, mechanical stores speak to the primary piece of deteriorating augmentations. Because of the pattern that reused materials are progressively brought into assembling forms, which were initially intended for new materials just, a proficient filtration turns out to be increasingly vital. In contrast to polyester, reused polyester utilizes PET as the crude material. This is a similar material that is utilized in clear plastic water containers, and recycling it to make the texture keeps it from going to landfill.

Material recycling is a procedure that gives business, helps philanthropy, maintains a strategic distance from the substantial ecological expense of sending the strong squanders to landfills and gathers dress for the penniless. Material squanders, for example, from fibre, material and dress can be sourced from the network, producing industry and buyers. These are otherwise called pre-buyer, post-customer and modern material waste [7]. In the meantime, post-buyer materials, regularly known as " messy waste", are gathered blended with other family unit things including both normal and manufactured materials, for example, fleeces, silks, rayon, woven nylon, cotton, polyesters furthermore, other material mixes. The recycling of waste from material saves assets and lessen the utilization of landfills. Besides, the purposes behind the open's readiness to partake and monetary market selection in recycling ought to be considered, as open cooperation is fundamental for viable recycling. In addition, the achievement of the market relies upon the clients' solicitations for reused merchandise and the mindfulness dimension of the network.

3   Pros in Recycled Polyester

Reused polyester gives a second life to a material that is not biodegradable and would somehow or another end up in landfill or the sea. Plastic has been found in 60% all things considered and 100% of all ocean turtle species since they botch plastic for nourishment, likewise as indicated by Ocean Conservancy. Reused polyester is nearly equivalent to virgin polyester as far as quality; however, its creation requires 59% less vitality contrasted with virgin polyester, as per a recent report by the Swiss Federal Office for the Environment. Furthermore, reused polyester can add to diminish the extraction of raw petroleum and gaseous petrol from the Earth to make increasingly plastic. Patagonia, outside brand says "Utilizing reused polyester decreases our reliance on oil as a wellspring of crude materials", best known for making downy from utilized soft drink bottles, unusable assembling waste and destroyed pieces of clothing. It likewise includes "It controls disposes of, accordingly dragging out landfill life and diminishing dangerous discharges from incinerators. It likewise advances new recycling streams for polyester attire that is never again wearable". Another American brand Nau says, "In light of the fact that polyester represents roughly 60% of the world's generation of PET—about twice what's utilized in plastic containers—building up a non-virgin production network for polyester fibre can possibly greatly sway worldwide vitality and asset necessities".

4 Cons of Recycled Polyester

Numerous pieces of clothing are not produced using polyester alone, but instead a mix of polyester and different materials. Overall, it is progressively troublesome, if certainly feasible, to reuse them. Now and again, it is in fact conceivable, for instance mixes with polyester and cotton. In any case, it is still at the pilot level. Certain overlays and completing connected to the textures can likewise render them unrecyclable. Indeed, even garments that are 100% polyester cannot be reused until the end of time. There are two different ways to reuse PET: precisely and artificially. Most rPET is gotten through mechanical recycling, as it is the less expensive of the two procedures and it requires no synthetic substances other than the cleansers expected to clean the info materials. Nevertheless, through this mechanical procedure, the fibre can lose its quality and therefore should be blended with virgin fibre. The vast majority trust that plastics can be interminably reused, yet each time plastic is warmed, it degenerates, so the consequent cycle of the polymer is debased and the plastic must be utilized to make lower quality items. The polyester chips produced by mechanical recycling can change in shading: some turn out firm white, while others are velvety yellow, making shading consistency hard to accomplish [8].

5 Dialog on the Results for the Textile Waste Recycling Strategies

The negative net qualities for practically all cases are from the kept away from effects coming about because of the framework extensions in each process, given that elective strategies for creating an equal measure of item, warmth, or power from essential assets are incorporated inside the framework limits. The decision to play out a framework development is plainly an imperative methodological decision since it overwhelms the all-out aftereffects of the frameworks. Given that this LCA is performed at a beginning period in the advancement of compound material recycling, accessible information fundamentally identify with vitality utilization. GWP and essential vitality use markers were chosen to be evaluated in this investigation to have a steady informational index. In any case, these markers do not secure other possibly essential natural effects [9].

The cellulose/polyester process is thought to be like the procedure of delivering fibre from wood-based assets; thusly, the nature of the reused yarns is thought to be as high given that the yarns are delivered from essential assets [10]. A modern leader is probably not going to confront a decision between precisely these arrangements of elective recycling systems given that the methods target distinctive pieces of the material waste stream and an immediate correlation may in this way not be applicable. Further, a few information are still very dubious. In any case, accepting that numbers are of the correct requests of greatness, in light of the introduced outcomes, for both essential vitality utilization and GWP, the vitality recuperation process, which is present practice, is the most exceedingly awful choice. Among the displayed recycling systems, material reuse is by all accounts the best choice attributable to the way that the assembling of items from essential assets is maintained a strategic distance from.

5.1 Patterns on Material Recycling

The natural effects of material creation, utilization and transfer are progressively considered in societal, business, research discusses, with differing degrees of information, and sway estimations gave. About the generation phases of the material esteem chain, consideration has been given to natural effects coming about because of cotton agribusiness, for example, biodiversity misfortune because of the utilization of pesticides and other harmful substances just as water exhaustion. While tanning and colouring forms antagonistically sway soil, water and air frameworks because of their arrival of lethal toxins, for example, chlorinated phenols and chromium, the completing and covering will in general include perilous substances affecting people and natural wellbeing, for instance in the utilization of fire resistant and water repellent wraps up.

6 Case Study on Recycling PET Bottles

6.1 PET Recycling

The POSTC-PET recycling industry began because of ecological strain to improve squander the executives. The other angle that goes about as main thrust for PET recycling industry is that PET items have a moderate rate of normal decay. PET is a non-degradable plastic in typical conditions as there is no known life form that can devour its generally expansive atoms. Figure 2 shows that PET Recycling process.

Confused and costly techniques should be worked with the goal for PET to debase organically. Recycling forms are the ideal approach to financially lessen PET waste. Then again, as the cost of virgin PET stays steady, new and less expensive advances for

Fig. 2 PET recycling

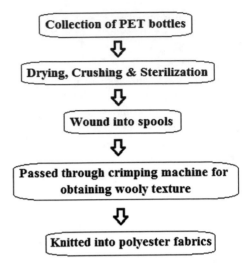

recycling PET give an additional incentive to the PET recycling industry by furnishing industry with generally less expensive PET. Numerous scientists announced that to accomplish fruitful PET recycling, PET chips should meet certain base necessities that are recorded in Table 1.

Notwithstanding the natural impetus, recycling of PET as an industry is getting its main thrust from the expanding worth and utilizations of virgin and altered PET. The most good working conditions that have been found in research centre instruments including chain extender fixation, living arrangement time, instrument type, and response temperature and rate speak to the learning base for building up the PET recycling further towards plausibility. Moreover, data on altered PET and chain extenders attributes is of incredible preferred standpoint to PET recyclers empowering them to choose best synthetic concoctions, conditions and hardware to suite their goals [11].

Table 1 Basic necessities of PET chips

Requirements	Value
Water content	<0.02 wt%
PVC content	<50 ppm
Flake size	0.4 mm < D < 8 mm
Metal	<3 ppm
Dye	<10 ppm
T_m	>240 °C
Polyolefin	<10 ppm

(i) **Points of interest to recycled polyester**

- Using more reused polyester lessens our reliance on oil as the crude material for our texture needs.
- Diverting PET containers for this procedure decreases landfill, and in this way lessens soil defilement and air and water contamination also, requires less vitality than virgin polyester.
- Garments made from reused polyester intend to be constantly reused with no debasement of value, enabling us to limit wastage. This implies polyester piece of clothing production could possibly turn into a shut circle framework.

(ii) **Difficulties and efforts**

- Solving quality issues through defilements of different added substances like cancer prevention agents, colours, stabilizers or hostile to blocking operator and shortening of the polymer chain at de-polymerization arrange.
- Finding substitutions for antimony, a polyester impetus known to be malignant growth causing (conceivably 500 mg/kg PET).
- Ensuring steady modern feedstock and shutting the circle by cultivating feedstock from material pre-and post-customer squander.
- Achieving recognizability and straightforwardness in the gathering, arranging and handling with social and reasonable conditions.
- Looking at lifecycle contemplations: biodegradability and recyclability of polymers.
- Transition towards inexhaustible biogenic feedstock transforming into mechanical strands.

(iii) **Advantages of recycled polyester**

- Spares characteristic assets
- Spares vitality underway procedure
- Lower ozone harming substance discharges and utilization of synthetic substances

While manufactured apparel when all is said in done is seen by numerous individuals as having a less regular feel contrasted with textures woven from characteristic filaments, (for example, cotton and fleece), polyester textures can give explicit points of interest over normal textures, for example, improved wrinkle obstruction, toughness and high shading maintenance. Therefore, polyester strands are in some cases spun together with common filaments to create a material with mixed properties. Engineered filaments additionally can make materials with predominant water, wind and natural opposition contrasted with plant-determined strands.

Pre-customer material squanders are those squanders which never make it to the shoppers and which come straightforwardly from the first producers. Precedents include: ginning squanders, opening squanders, checking squanders, comber noils, brushed squander yarns, wandering squanders, ring turning waste strands, ring spun squander yarns, open end turning waste filaments, open end turning yarn squanders, sewing waste yarns, weaving waste yarns, texture cutting squanders, texture wet preparing squanders, clothing producing squanders, etc. Post-mechanical material squanders are created amid the assembling procedure of upstream items and regularly they are from the virgin fibre makers, tire rope producers, polymerisation plants, and other plastic items. Post-shopper material squanders are the squanders that are recuperated from the shopper inventory network and these are commonly the garments that are prepared for transfer or landfill. Well known models incorporate reusing of the extras and drink jugs to make reused polyester.

6.2   The Issue of Shaded Waste Streams

As per the organization, one specific advantage of the innovation is its capacity to effectively manage shaded containers. Current polyester recycling techniques available require a generally perfect waste stream of clear or light blue polyester bottles. Therefore, it calls attention to; it is not frequently monetarily practical to reuse shaded or blended polyesters. This implies a great deal of shaded plastic waste will even now end up in landfill or cremation. "It empowers a persistent polyester recycling process for different defiled polyester squander streams—including bundling materials and material items—and cleans the first plastic material of contaminants and shading."

6.3   Polyester Recycling Word Wide—Advancement and Viewpoint

(i) **USA**

As wilderness of numerous mechanical advancements, the USA began the primary considerable PET recycling amid the late 70th. Low crude material expense—around then utilized PET jugs where junk—and ecological exercises of an exceptionally created society have been the significant main thrusts to create inside a generally brief time a US polyester recycling industry. The immersion of this advancement came amid the late 90th and the primary long periods of the new century. The accomplished gathering dimension of 20–25% ended up dormant for a few reasons. Principally determined by the forcefully rising costs of baled bottles and the general down pattern of the US-material industry the enthusiasm for PET-recycling contracted. Driving drive behind was the detonating material industry in China associated with a gigantic cost increment of gathered/baled PET-bottles.

Today for the most part the brand proprietors in USA like Coca Cola keeping the PET recycling running on a dissatisfactory dimension by driving the jug maker to apply a fix measure of 10–30% of reused polymer in each new pitcher.

(ii) **Europe**

Very unique the advancement in Europe. After a moderate begin in Europe amid 90th other instruments came in power. Germany like the green spot just as comparative EU guidelines. Together with a "green–natural" driven governmental issues and comparative developments in advertising Europe was updating the USA in accumulation rate and the aggregate sum of gathered PET pitcher.

The EU garments and materials part involved 185,000 organizations, which utilized 1.7 million specialists. Organizations with under 50 representatives represented over 90% of the workforce and delivered practically 60% of the esteem included, demonstrating that the segment is based around private companies. The EU business has seen radical changes in which a few organizations have kept up and reinforced aggressiveness by diminishing large scale manufacturing and focusing on a more extensive assortment of higher esteem included items, through quality, plan and mechanical development. To be sure the top of the line division became quicker than the remainder of the European economy amid the ongoing monetary emergency, utilizing more than 1 million individuals, sending out over 60% of creation outside Europe, and representing 10% of all EU trades.

The general picture of the division as a low-benefit, low development industry with low wages and flawed working conditions has immensely decreased its appeal to youthful labourers and experts, making the enlistment of a talented workforce risky. Added to this are troubles discovering credit, districts where business enterprise is immature and an absence of effectively recognizable best practice endeavour precedents.

(iii) **France**

In France, material materials that are gathered independently are largely reused through abroad fares of second-hand garments (60%), while around 32% are destroyed and down cycled into nonwovens, protection felts and cleaning wipes. Around 7.5% is burned to deliver vitality.

(iv) **Belgium**

Belgium has truly practically identical figures, despite the fact that the level of fibre reuse through destroying is higher (47%). As featured by these figures, the accumulation of material waste stays ruled by the mission for high quality and reusable garments. This reuse part is trailed by recycling exercises, ordinarily by methods for mechanical procedures that are considered "down cycling" because of the loss of fibre quality and quality. The gatherers' calling is subsequently centred on item instead of material. This can be clarified by the way that the innovations to sort, independent and procedure solitary fibre type are still at the improvement organize, and not yet up scaled to modern practice.

(v) **China**

Chinese sources are guaranteeing an accumulation rate of over 90%! In China in view of countless poor vagrant specialists PET suppresses are picked wherever they are dropped. The Chinese fibre producers figured out how to deal with and process bottle drops. Together with a noteworthy deficiency of PTA and PET creation when all is said in done amid 2003–2005 time the Chinese began everywhere scale to import the pieces from everywhere throughout the World. Informal information are checking the change measure of PET jug pieces in China 2006 to around 2000 kt/a.

(vi) **India**

In India, the board of plastic waste is a major test, which for the most part establishes PET and PE squander, subsequently this examination considered these two noteworthy plastic squanders. The essential goal of this investigation is to assess the effects on nature because of different plastic waste administration choices. This examination additionally considered accumulation and transportation period of PET and PE squander recycling and incorporated its effects in all the four situations. The aftereffects of this investigation will demonstrate the effects on the earth for accumulation and transportation of PET and PE squander and the four plastic waste administration situations (landfilling, cremation without vitality recuperation, recycling of PET and PE waste and burning with vitality recuperation).

(vii) **Germany**

This examination puts critical spotlight available conditions in the district with the mean to recognize approaches to increment the interest of reused plastics in household industry, and in a perfect world accomplish this by activating territorial assets. Germany as an essential outside on-screen character in the recycling execution of the nations in this investigation, with generally 40% of plastic waste being sent out from the Scandinavian locale every year for recycling in Germany Stress that the examination centers around post-shopper plastic waste, both from family units and business, and does not think about the administration of modern waste which is overseen under various conditions and includes less moderate on-screen characters, in this manner speaking to a less difficult esteem chain. Besides, a noteworthy portion of modern waste is dealt with inside a similar industry or through cross-division cooperation under modern beneficial interaction, hence not entering the recycling market.

(viii) **Finland**

Recycling of plastics is asked by the requirement for shutting material circles to keep up our characteristic assets when endeavoring towards round economy, yet in addition by the worry dashed by perceptions of plastic piece in seas and lakes. Bundling industry is the area utilizing the biggest offer of plastics, subsequently bundling overwhelms in the plastic waste stream. In Finland post-purchaser plastic bundling is gathered either independently in the bring site accumulation in the blended metropolitan strong waste (MSW) or in vitality squander. Bring site accumulation with provincial gathering directs close toward business sectors and stores, and so on was begun in 2016 because of the Government order on bundling and bundling waste, which stipulated the makers full obligation of bundling waste administration. Likewise, a few regions have propelled post-customer plastic bundling accumulation preliminaries, which offer a less complex route for buyers to sort and dispose of their plastic bundling squanders near, and dear, at properties. Separate gathering is expanding bit by bit, and procedures for the arranging, washing and recycling of plastic bundling were propelled as of late. To survey the recycling capability of post-shopper plastic bundling waste in Finland, two kinds of waste streams were contemplated, in particular plastic bundling gathered as a component of blended MSW and plastic bundling gathered independently.

(ix) **African nations**

Zambia like numerous African countries faces the test of overseeing strong waste. The poor waste administration has added to poor sanitation and wellbeing related issues. In spite of strong waste administration related difficulties, Zambia has squander strategies and enactments set up. In view of guidelines and laws controlling the executives of strong waste exist however there are not kidding implementation and consistence insufficiencies. Accordingly, as a aftereffect of proceeded with endeavors to improve the waste administration segment in Zambia, the national strong waste administration system was created in 2004 and one of its destinations is squander minimization and recycling. Nevertheless, the development of this system has not completely prompt the recuperation and recycling of plastic strong waste from the waste streams just as from generators. The difficulties in the recuperation and recycling of PSW ought not to depend on the approaches and enactments set up in request be tackled. Plastic assembling and recycling organizations ought to evaluate the effect of actualizing other procedures to accomplish maintainable asset recuperation and recycling.

7   Global Concern

(i) **Techniques for sustainable administration of plastic solid waste (PSW)**

The way that plastics are fabricated from restricted asset, a ton of innovative advancements for recycling plastics among different assets are been made. As a worldwide issue, more ought to be done to explain this present world's asset and procedures other than innovation are required [12, 13]. Various procedures as, financial, natural, social, and market exist. These techniques are cardinal in the advancement of frameworks for the recuperation and recycling of PSW [14]. To economically fabricate recuperated and reused PSW it is vital to comprehend if the procedure will be monetarily, earth, legitimately and socially adequate to embrace. In this way, the discoveries from this exploration will add to the group of learning on practical assembling of PSW as well as waste administration by featuring the basic variables for execution in recuperation and recycling programs for this waste sort [15].

The Global Recycle Standard (GRS), started by Control Union and now directed by Textile Exchange is expected to build up autonomously confirmed cases with regards to the measure of reused content in a yarn, with the critical included component of precluding certain synthetic compounds, requiring water treatment and maintaining labourers rights, holding the weaver to models like those found in the Global Organic Textile Standard. There is yet an issue with the generation of synthetics. Prospering proof about the tragic results of utilizing plastic in our condition keeps on mounting. Another accumulation of friend-audited articles, speaking to more than 60 researchers from around the globe, expects to evaluate the effect of plastics on the earth and human wellbeing. In landfills, they discharge overwhelming metals, including antimony, and different added substances into soil and groundwater. On the off chance that they are scorched for vitality, the synthetic substances are discharged into the air. On the off chance that you pick an engineered, at that point you sidestep the advantages you get from supporting natural horticulture, which might be a standout amongst our most powerful weapons in battling environmental change, in light of the fact that:

- Natural farming goes about as a carbon sink: new research has demonstrated that what is IN the dirt itself (microorganisms and other soil life forms in solid soil) is progressively imperative in sequestering carbon that what develops ON the dirt. What's more, contrasted with backwoods, rural soils might be an increasingly secure sink for air carbon, since they are not powerless against logging and out of control fire. The Rodale Institute Farming Systems Trial (FST) soil cell based information (which covers 30 years) exhibits that improved worldwide earthbound stewardship—explicitly including regenerative natural horticultural practices—can be the best presently accessible procedure for alleviating CO_2 emanations.
- It disposes of the utilization of engineered manures, pesticides and hereditarily adjusted living beings (GMOs) which is an improvement in human wellbeing and agro biodiversity.
- It saves water (making the dirt progressively friable so water is retained better—decreasing water system necessities and disintegration).
- It guarantees continued biodiversity.

The greatest downside to polyester generation is that it requires a ton of vitality, which implies consuming fuel for power and adding to environmental change. Anyway, industrial facilities where polyester is created which don't have end-of-pipe wastewater treatment frameworks discharge antimony alongside a large group of other conceivably risky substances like cobalt, manganese salts, sodium bromide, and titanium dioxide into the earth. In principle, cotton is biodegradable and polyester is not. Nevertheless, the thing is the manner in which we discard dress makes that insignificant. For cotton garments to separate, they must be treated the soil, which does not occur in a landfill.

Recycling of polyester is advancing to coordinate the prerequisites of the necessities. Mechanical and synthetic reusing of polyester are the two noteworthy advancements right now accessible. The innate faults of mechanical reusing limit the end utilizes/applications somewhat, yet artificially reused polyester is discovering its applications in a wide assortment of items. Generally, a wide range of employments are being produced from reused polyester. Refreshment bottles delivered from reused PET are on the ascent. Purchaser acknowledgment will be a noteworthy standard in making polyester reusing a genuine maintainable.

References

1. Cuc S, Iordanescu M, Girneața A, Irinel M (2015) Environmental and socioeconomic sustainability through textile recycling. Ind Text 66(3):156–163
2. Shigemoto I, Kawakami T, Taiko H, Okumura M (2011) A quantum chemical study on the polycondensation reaction of polyesters: the mechanism of catalysis in the polycondensation reaction. Polymer 52:3443–3450
3. Biresaw G, Carriere CJ (2004) Compatibility and mechanical properties of blends of polystyrene with biodegradable polyesters. Compos A Appl Sci Manuf 35:313–320
4. Yousef S, Tatariants M, Tichonovas M, Sarwar Z, Jonuskiene I, Kliucininkas L (2019) A new strategy for using textile waste as a sustainable source of recovered cotton. Resour Conserv Recycl 145:359–369
5. Arturi KR, Sokoli HU, Sogaard EG, Vogel F, Bjelic S (2018) Recovery of value-added chemicals by solvolysis of unsaturated polyester resin. J Clean Prod 170:131–136
6. Shokrieh MM, Rezvani S, Mosalmani R (2017) Mechanical behavior of polyester polymer concrete under low strain rate loading conditions. Polym Test 63:596–604
7. Liu W, Liu S, Liu T, Liu T, Zhang J, Liu H (2019) Eco-friendly post-consumer cotton waste recycling for regenerated cellulose fibers. Carbohyd Polym 206:141–148
8. Pagliaro M (2017) Chapter 3—Esters, ethers, polyglycerols, and polyesters, glycerol, pp 59–90
9. Tang X, Chen EY-X (2019) Toward infinitely recyclable plastics derived from renewable cyclic esters. Chem 5:284–312
10. Lavoratti A, Scienza LC, Zattera AJ (2016) Dynamic-mechanical and thermomechanical properties of cellulose nanofiber/polyester resin composites. Carbohyd Polym 136:955–963
11. Wypych, G., (2012) PET poly(ethylene terephthalate). Handbook of polymers, pp 385–390
12. Mwanza BG, Mbohwa C, Telukdarie A (2018) Strategies for the recovery and recycling of plastic solid waste (PSW): a focus on plastic manufacturing companies. Proc Manuf 21:686–693
13. Delva L, Hubo S, Cardon L, Ragaert K (2018) On the role of flame retardants in mechanical recycling of solid plastic waste. Waste Manag 82:198–206
14. Mwanza BG, Mbohwa C (2017) Drivers to sustainable plastic solid waste recycling: a review. Proc Manuf 8:649–656
15. Singh N, Hui D, Singh R, Ahuja IPS, Feo L, Fraternali F (2017) Recycling of plastic solid waste: a state of art review and future applications. Compos B Eng 115:409–422